THE
NORTH
WESSEX
DOWNS

In this series

A Portrait of the Surrey Hills
The Natural Beauty of Cornwall
Chichester Harbour: England's Coastal Gem

THE NORTH WESSEX DOWNS

Steve Davison

ROBERT HALE · LONDON

© Steve Davison 2013
First published in Great Britain 2013

ISBN 978-0-7198-0581-3

Robert Hale Limited
Clerkenwell House
Clerkenwell Green
London EC1R 0HT

www.halebooks.com

The right of Steve Davison to be identified as
author of this work has been asserted by him
in accordance with the Copyright, Designs and
Patents Act 1988

A catalogue record for this book is available from the British Library

10 9 8 7 6 5 4 3 2 1

Typeset by Dave Jones
Printed in China

Contents

Acknowledgements 7

Introduction 8

1. Geology and the landscape 12

    Underlying Geology 12

    The Landscape 17

    Chalk Streams and Rivers 23

2. History – from the Stone Age to the present day 27

    A Potted History 27

    Archaeology Uncovered – from Causewayed Camps to
        Medieval Castles 31

    North Wessex at War 51

3. North Wessex at work 59

    From the Land to Science 60

    Boats, Trains and Cars 71

    Tourism 79

4. Towns and villages 82

    Main Market Towns 82

    Kennet Valley Villages 89

    Pang and Thames Valley Villages 93

    Villages in the Berkshire Downs 98

    Villages in the Northern Part of the AONB 101

    Villages in the Lambourn Valley Area 103

    Villages to the South of Swindon 105

    The Vale of Pewsey and the River Bourne Valley 106

    Along the North Hampshire Downs 108

    Hampshire Villages in the Bourne Rivulet Valley 111

5. From stately homes to white horses 114

    Stately Homes 114

    Interesting Churches 124

    Monuments, White Horses and Crop Circles 137

6. Writers and artists 145

    The Bloomsbury Set 145

    Writers of Prose and Poetry 145

    Artists of Land and Light 154

7. Wildlife 156

    The Flora and Fauna 156

    From National Nature Reserves to Wildlife Trust Reserves . . . 159

    More Places to Visit 165

Useful contacts and information 167

Glossary 170

Further reading 172

Index 174

Acknowledgements

The author would like to thank the North Wessex Downs AONB team for providing information about the North Wessex Downs, also the Berkshire, Buckinghamshire and Oxfordshire Wildlife Trust (especially Wendy Tobitt), the Hampshire and Isle of Wight Wildlife Trust and the Wiltshire Wildlife Trust for their input regarding their reserves within the AONB. Thanks also go to members of the RSPB North Wessex Farmland Bird Project team, Lesley Dunlop from the Berkshire Geoconservation Group, various local history groups and everyone else who took time to answer my requests for information.

Introduction

Throughout England, Wales and Northern Ireland there are forty-six Areas of Outstanding Natural Beauty (AONB) covering around 18 per cent of the British countryside, areas of precious landscape whose distinctive character and natural beauty need to be safeguarded for future generations. The thirty-three AONBs that lie wholly within England have been described as the *'jewels of the English landscape'*.

One of these areas, designated in 1972, is the North Wessex Downs AONB. Covering an area of 668 sq miles, the North Wessex Downs is the third largest AONB after the Cotswolds and the North Pennines.

The area forms part of the Southern England Chalk Formation, which runs from Dorset in the west to Kent in the east and also includes the Dorset Downs, Purbeck Hills, Cranborne Chase, Wiltshire Downs, Salisbury

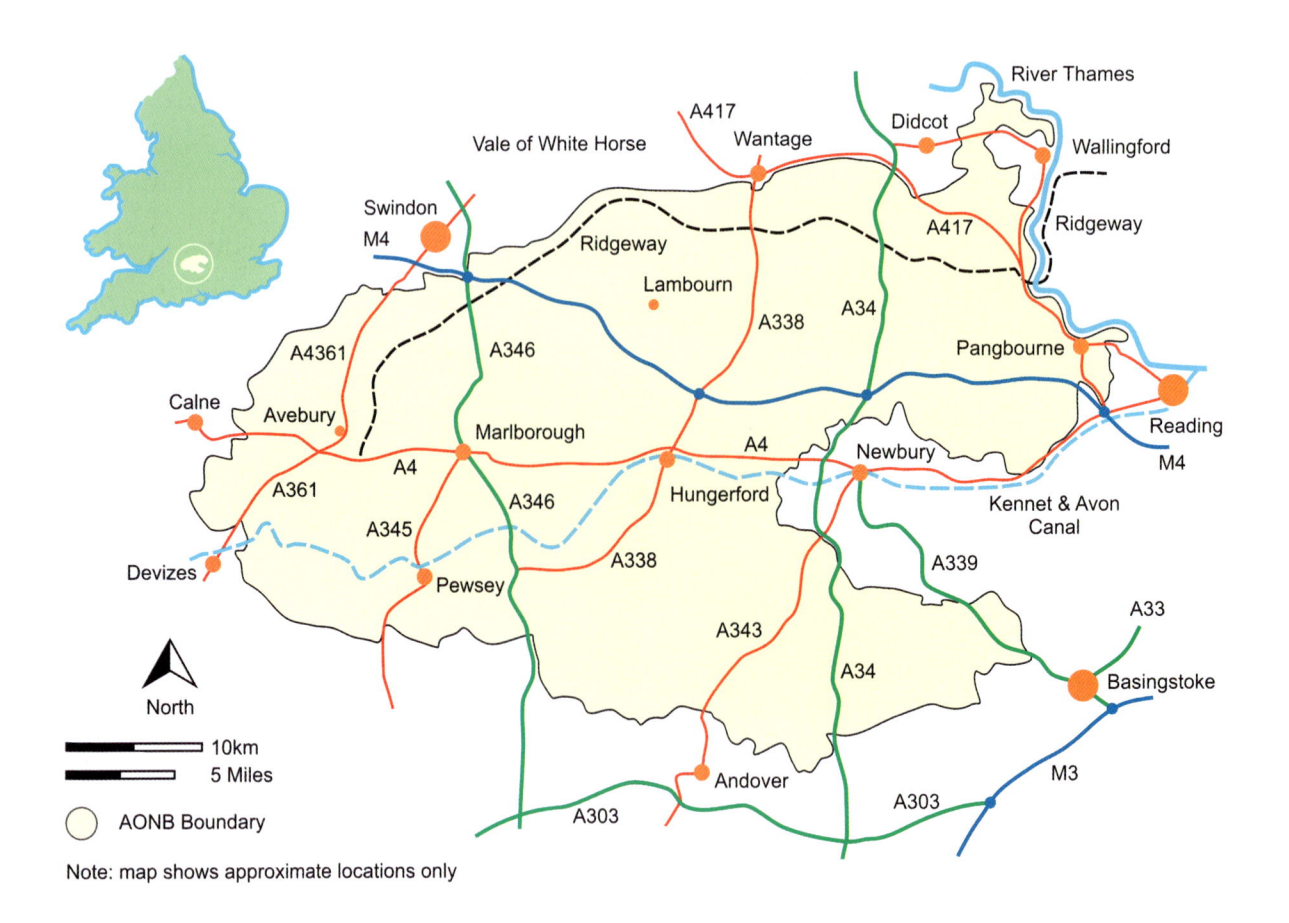

▲ North Wessex Downs Area of Outstanding Natural Beauty

INTRODUCTION

▲ Looking north across the Kennet Valley from Inkpen Hill in the North Hampshire Downs

▲ The River Thames, near Pangbourne, forms the eastern edge of the North Wessex Downs, with the Chilterns AONB across the river

▲ The River Kennet: one of the many lovely chalk streams to be found within the North Wessex Downs (seen here near Mildenhall)

Plain, the Isle of Wight, Chiltern Hills and the North and South Downs.

The North Wessex Downs stretch from their western tip near Calne in Wiltshire across a broad arc to the south of Swindon, passing through Berkshire and Oxfordshire, with a steep scarp slope looking out over the Vale of White Horse to meet the River Thames on its eastern edge, adjoining the Chilterns AONB across the Goring Gap. Along the crest of the downs, followed for much of the way by the Ridgeway – probably the oldest green road in England – prehistoric man has left us some fascinating treasures including the beautiful Uffington White Horse, the magical Wayland's Smithy and several Iron Age hill forts with commanding views.

The AONB then sweeps south, following the River Thames to Pangbourne before encircling Newbury and part of the Kennet Valley, to encompass the northern reaches of the Hampshire Downs, home to Watership Down (known the world over as the home of a colony of rabbits in Richard Adams' *Watership Down*) and Walbury Hill, the highest chalk hill in England. The southern edge stretches westwards, passing north of Basingstoke and Andover to take in the Vale of Pewsey, enclosing the beech- and oak-wooded Savernake Forest and the market towns of Hungerford and Marlborough, before heading back towards Calne.

The predominant feature is the underlying chalk geology; the North Wessex Downs cover one of the most continuous tracts of chalk downland in England. The chalk itself is formed from the remains of billions of minute sea creatures – known as coccoliths – that over time have been compacted and raised above sea level by the gradual movement of land masses drifting across the earth's surface. Yet the North Wessex Downs' landscape is one of great diversity; not only are there the high open chalk downlands incised by narrow chalk river valleys, but there are more intimate, well-wooded areas such as the forests of Chute and Savernake, and in parts where the land is lower there is a rich mosaic of woodland, pasture, heath and common land. The poet Edward Thomas (1878–1917) described the region in the following terms: 'This is pure downland, the breasted hills curved as if under the influence of a great melody. It is beautiful, a quiet, and unrenowned and a most visibly ancient landscape.'

Although the chalk downland and ancient woodland are of national importance, the landscape has been etched by the impact of humans for over 5,000 years. The archaeology of the North Wessex Downs is both rich and varied, with a number of impressive monuments including the Neolithic stone circle at Avebury, the West Kennet Long Barrow and mystical Silbury Hill (all of which forms part of a World Heritage Site), the sites along the Ridgeway and a myriad of Bronze Age barrows and Iron Age hill forts.

The name North Wessex Downs is not a traditional one, although Wessex was indeed a region of Saxon Britain. The name Wessex was later revived by Thomas Hardy in his novels, and its northern reaches did include much of the present North Wessex Downs, so maybe it is a fitting and evocative name for one of the largest tracts of chalk downland in southern England. As for the word 'downs', we need to thank the Anglo-Saxons for it; their word for hill was '*dun*'.

Although located in southern England – a densely populated region – the area of the North Wessex Downs is in fact sparsely populated, with a population at the time of writing of just 125,000, mostly concentrated in the valleys and the market towns of Marlborough and Hungerford, leaving large areas of virtually uninhabited downland with a true sense of isolation. Mile after mile of footpaths offer breathtaking views, while hidden amongst the folded contours are picture-postcard villages with thatched cottages, historic churches and cosy pubs.

While the major land use of the North Wessex Downs is agriculture, and racehorse training is an important industry (especially in the Lambourn Valley, which is known as the Valley of the Racehorse), the region has an equally important recreational role for both visitors and the immediate population.

The underlying geology and topography, cultural evolution and land management of the North Wessex Downs combine to offer much of interest to those who wish to explore the region. It is an ancient landscape with a great sense of remoteness and tranquillity that has evolved through human impact for millennia … let our own exploration begin across the length and breadth of this magical piece of 'chalk country'.

A Note on Finding some of the Locations

In various parts of this book, six-figure grid references have been provided so that individual sites can be located on Ordnance Survey maps; a list of maps covering the AONB is provided in Useful Contacts and Information, towards the end of this book. Each map is divided by a series of vertical and horizontal lines to create a grid with a spacing of 1km. You can locate a point on a map, accurate to within 100m, using a grid reference which is made up of two letters and six numbers. The two letters correspond to the 100,000 metre square in which the grid reference lies. The first two numbers correspond to the vertical running line (known as 'eastings') to the left of the point of interest, using the horizontal numbers along the bottom and top of the map; the third number is the tenths of the square (equivalent to 100m). Next, take the fourth and fifth numbers, and move up the map to locate the horizontal line ('northings') below the point of interest; the last digit is again the number of tenths moving up through the square. Always remember – the horizontal numbers come before the vertical ones.

CHAPTER ONE

Geology and the landscape

Underlying Geology

Before exploring the treasures of the North Wessex Downs we should spend some time having a look at the underlying geology, as this gives rise to many of the region's special characteristics.

The geology of the region tells the story of the seas that once covered southern England and the sediments that were laid down at that time. It is perhaps easiest to think of the geological structure as a multi-layered cake. The lowest layers are formed of the oldest rocks from the Lower Cretaceous period. In the middle is a thick layer of Upper Cretaceous chalk and at the top are the Palaeogene deposits.

LOWER CRETACEOUS (145–99 MILLION YEARS AGO)

These are the oldest rocks in the region, although only the topmost layers of the Lower Cretaceous – the Gault Clay and Upper Greensand – are found within the North Wessex Downs. The Gault Clay, a blue-grey mudstone, is restricted to a narrow band along the northern foot of the downs below the steep scarp slope. This layer of clay has historically been used for brick-making, especially around Swindon and Devizes (the band of Gault Clay continues westwards along the northern edge of Salisbury Plain).

Above this are the silts and sands of the Upper Greensand, which are rich in a mineral called glauconite

▲ Classic chalk downland scenery (North Hampshire Downs near Cottington Hill)

that gives the sands their green tinge. This layer of Upper Greensand follows the northern scarp of the downs, forming a junction with the underlying Gault Clay. It is this junction between the impervious clay and the more porous layers above that gives rise to a spring-line along the northern edge of the downs. The Upper Greensand is also seen on the surface along the Vale of Pewsey as a result of folding in the geological layers. It is additionally seen around Kingsclere.

UPPER CRETACEOUS (99–65 MILLION YEARS AGO)

Overlying the Lower Cretaceous rocks is a 700 foot-thick layer of chalk which dominates both the geology and landscape; across southern England this layer of chalk varies in thickness from around 700 to 1,500 feet or more, with some of the variation having been caused by erosion over millions of years. Chalk, a special type of soft white limestone, is formed from incredible numbers of minute calcareous shells which are the remains of plankton, known as coccoliths, which lived in shallow subtropical seas that once covered much of southern England.

This chalk is split into three layers, namely Lower, Middle and Upper (with the Lower layer being the oldest and the Upper layer the most recent). Due to both plate tectonics (the gradual movement of continents across the surface of the earth that caused the layers of chalk to ripple) and erosion, these three bands of chalk are 'exposed' at different locations throughout the North Wessex Downs. The layer of Lower Chalk,

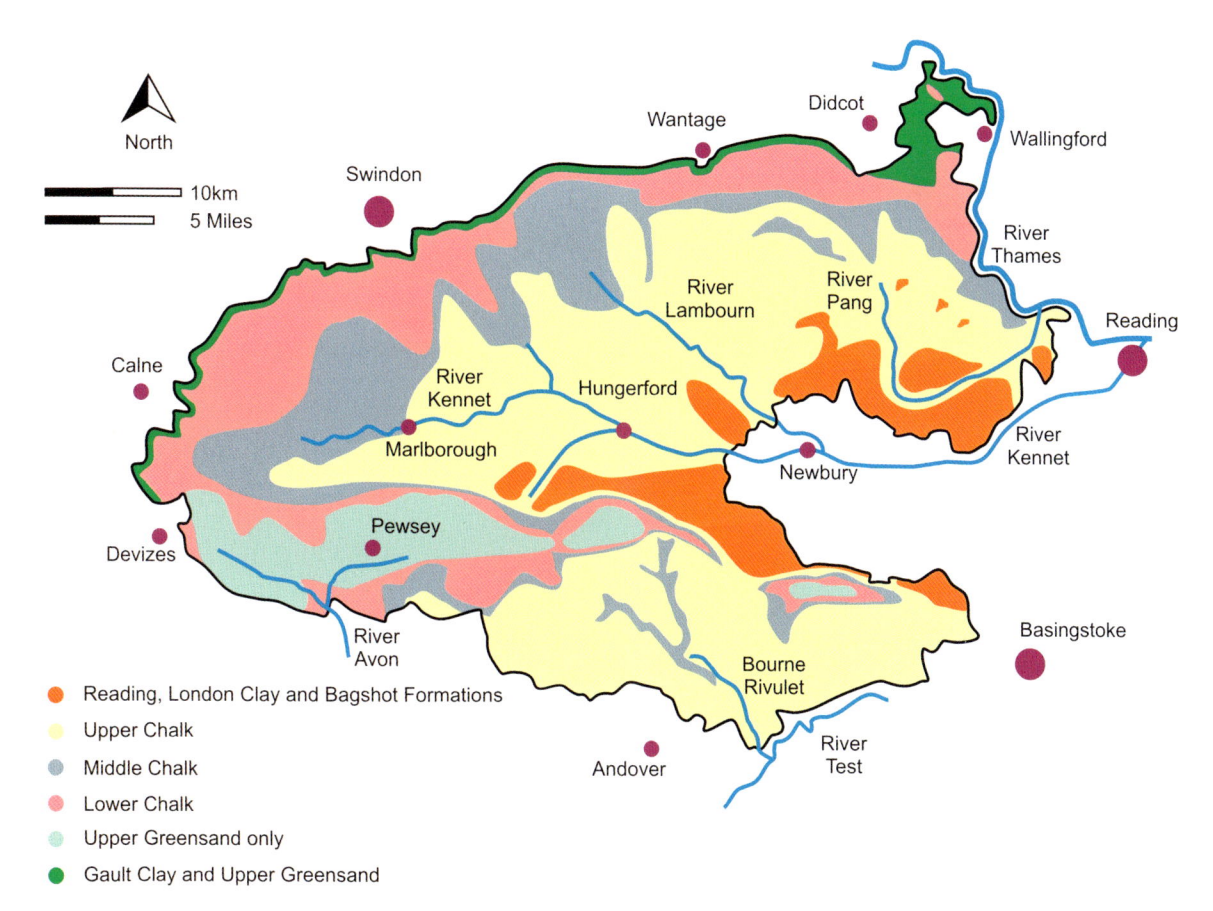

▲ Simplified map of the geology of the North Wessex Downs

which has a greyish colour due to the inclusion of clay, forms the dramatic scarp slope along the northern edge of the North Wessex Downs and also around the Vale of Pewsey and the North Hampshire Downs. This is closely followed by a 'ring' of Middle Chalk; however, the vast bulk of the downs consists of the Upper Chalk, which has a much whiter colour than the Lower Chalk.

Chalk is a highly porous rock with numerous microscopic pore spaces that can store huge amounts of water, thus it acts like a giant sponge, soaking up most of the rainfall. Because of this, over most of the higher ground there is no surface water in the form of ponds or streams. However, this characteristic does give rise to the celebrated chalk streams of the region.

Chalk stone (known as 'clunch') has sometimes been used as a building material such as in the spring-line villages of Bishopstone, Ashbury and Woolstone. However, chalk is not weatherproof and is easily eroded by rain, necessitating the need for both good foundations and overhanging thatched roofs to protect the walls, giving rise to the old saying 'good shoes and a hat'.

▲ Woolstone Church is made from chalk stone or clunch

▲ Upper Chalk with a band of flint

Associated mainly with the Upper Chalk are horizontal bands of irregular silica concretions, known as flints. The process by which these develop is not fully understood, but it is probably the result of high concentrations of silica dissolved from the skeletons of sponges and plankton that form the chalk. Flints also occur in profusion in the jumbled deposits of weathered chalk known as 'clay-with-flints'. When struck, flint breaks with a shell-shaped fracture, leaving very sharp edges, and our Stone Age ancestors used flints to make a range of tools such as knifes, scrapers, hand axes and arrowheads. Being a very hard-wearing rock, flint has been widely used as a building material both in its natural state, and knapped to form a flatter surface, and it is a characteristic of the area – a great number of the region's churches and old cottages have flint walls.

TERTIARY (65–2.6 MILLION YEARS AGO)

The deposits from the Tertiary and in particular the Palaeogene period from 65–23 million years ago – the Reading Formation, London Clay Formation and Bagshot Formation – are concentrated in an east–west line to the south of Hungerford and to the north of Thatcham, and were formed in shallow marine and fluvial (river) environments. The Reading Formation is mainly responsible for the sarsen stones, known locally as *grey wethers* – from a distance they are said to resemble sheep (a 'wether' being the term for a castrated ram), although the word sarsen is believed to be derived from 'saracen' meaning 'foreigner'. These blocks of hard sandstone are one of the most identifiable and well-known aspects of the North Wessex Downs.

The sarsen stones, isolated remnant blocks of Tertiary sand bound together with hard silica cement, were possibly formed around 10–5 million years ago, during arid conditions when the underlying water table was at a similar depth to the sedimentary deposits. The surrounding uncemented sediments have long been eroded away, leaving the much harder deposits as blocks lying on the surface.

The stones were used as far back as Neolithic times in the construction of the stone circle at Avebury and

▲ Picturesque cottage in Ashbury built from chalk and showing the typical 'good shoes and a hat' (overhanging thatched roof and a brick base to the walls)

▲ The Manger just below the famous Uffington White Horse was formed by water erosion during the last ice age

the Neolithic long barrows at West Kennet and Wayland's Smithy. More recently, sarsens have been broken up for use as a building material in walls and houses to such an extent that many of these stone deposits have virtually disappeared. In the early eighteenth century the English antiquarian William Stukeley saw at first-hand the destruction of many of the large sarsen stones around Avebury (see page 33).

It was because of rising concerns early in the twentieth century that before long few or no sarsens would be left that the Grey Wethers Preservation Fund was founded and resulted in the purchase of Piggledene (SU141686) and Lockeridge Dene (SU146674) for the National Trust. Sarsens can also be seen lying at the foot of Weathercock Hill and Kingstone Down near Ashdown House (SU285820). However, by far the largest concentration of these strange rocks is at Fyfield Down National Nature Reserve (NNR) where the ground is literally littered with sarsen stones of varying sizes. The strange landscape of Fyfield Down can easily be reached by following a signed path from a car park (SU159699) just off the A4 to the west of Marlborough. In 1668 Samuel Pepys visited the area around Fyfield Down and wrote: 'It was prodigious to see how full the Downes are of great stones; and all along the vallies, stones of considerable bigness, most of them growing certainly out of the ground so thick as to cover the ground.'

It was during the Tertiary period that the European and African continental plates collided – an event that formed the Alps. However, in southern England the effects were less dramatic, although the collision caused our sedimentary 'cake' to ripple and buckle, with consequent effects upon the landscape that remain in evidence today.

The exposed upper (northern) edge of the chalk layer forms a ridge which runs past Whitehorse Hill (856 feet) near Uffington to the Thames at Goring and then continues along the Chiltern ridge. This sheet of chalk passes south under the Kennet Valley and reappears to form a ridge – the North Hampshire Downs – which includes Walbury Hill (974 feet), the highest chalk hill in England; other high points in the North Wessex Downs include Milk Hill and Tan Hill near Alton Priors to the east of Devizes at 968 feet. In 2009 the Ordnance Survey confirmed that Tan Hill is some 10 inches higher than Milk Hill, making the former the highest point in Wiltshire.

This rippled layer of chalk also forms the North and South Downs (South Downs from Hampshire and through East and West Sussex; North Downs from the Hampshire/Surrey border through to the White Cliffs of Dover in Kent), and the Purbeck Hills in Dorset.

QUATERNARY DEPOSITS (2.6 MILLION YEARS AGO TO PRESENT)

Throughout the last 2.6 million years Britain has been subjected to several periods of glaciation separated by warmer interglacial periods (the last glacial period ended about 12,000 years ago). During these glacial periods much of Britain was hidden beneath a thick layer of ice, although there is no evidence to suggest that the North Wessex Downs, or the rest of southern England, were ever covered in ice. However, the area did suffer periglacial conditions which allowed the formation of dry valleys, or coombes, in the Cretaceous chalk plateau, such as The Manger below Whitehorse Hill and both Crowhole Bottom and the adjacent Devil's Punchbowl near Letcombe Bassett. These are the result of water flowing over the surface of the chalk during cold periods when the underlying ground has been frozen, making the normally porous chalk impermeable.

Another major feature caused by glaciation was the creation of the Goring Gap and the diversion of the Thames southwards to flow past Reading; originally the river flowed through the Vale of St Albans, past Watford and Hertford, eventually reaching the North Sea in East Anglia near Ipswich. The gap was created when a large glacial lake, which formed over the Oxford area about 450,000 years ago, eroded a line of weakness in the chalk. The Goring Gap now forms a junction between the Berkshire Downs to the west and the Chiltern Hills to the east.

Within the North Wessex Downs there are three principal types of Quaternary deposit: the river terrace deposits, alluvium or modern-day river sediments and a deposit known as clay-with-flints.

The river terrace sediments, which are formed predominantly of gravels with sand, are to be found along all the main river valleys within the AONB. These sediments were most likely deposited during the cold periods and not during the warmer interglacial periods of the Quaternary.

The development of the terrace sediments reflects the successive cutting of ancient floodplains; the older terraces being the highest ones. Alluvium is the modern-day deposits of rivers and is mostly formed of silt and clay, but can also contain sand and gravels during periods when the rivers are flooded.

Unlike terrace sediments and alluvium, the clay-with-flints is not the result of river activity. It is thought more likely to be the remains of the chalk after the effects of intense weathering.

The Landscape

Having looked at what lies beneath the surface of the North Wessex Downs we can now turn our attention to what lies above ground.

The landscape is characterized by rolling hills and vales with steep scarp slopes along the northern edges of the chalk. The open areas of chalk downland are devoid of surface water and have limited habitation. Valleys are commonly dry – having been cut by streams during the last Ice Age, when permafrost made the chalk impermeable. Now rainwater sinks into the porous rock until, at lower levels, the water table is reached and, where this reaches the surface (at the junction of the porous chalk and lower impervious clays), springs emerge.

This is typically an agricultural landscape with over 80 per cent of the area classed as some type of farmland, including wooded plantations. Throughout history our ancestors have continually changed and managed the landscape to fit their needs at the time. On top of this they have left a rich archaeological legacy from Neolithic long barrows to Iron Age hill forts, massive Saxon earthworks such as the Wansdyke, and former royal hunting forests.

Population growth in the thirteenth century led to a demand for more farmland, causing the fragmentation of the old hunting forests into small individual deer parks – still a characteristic of the eastern part of the North Wessex Downs. More intensive land use was introduced, with new areas being cleared and ploughed, as evidenced by the many strip lynchets or terraces still visible on the downland slopes.

More recently towns and villages have encroached onto the surrounding countryside and the demand for water from the chalk aquifer continues to threaten the ecosystem of the chalk stream – around 70 per cent of the public water supply in the south-east of England comes from the chalk aquifer. Transportation has changed from horse and cart, to canal boats, railways and an expanding road network. The landscape of the North Wessex Downs is most definitely a continually changing canvas.

Studies have shown that the landscape of the region can be split into eight distinct landscape character types – let us take a brief look at these different land types.

DOWNS PLAIN AND SCARP

Extending along the entire length of the northern boundary of the North Wessex Downs, from Cherhill in the west to Chilton in the east, with an outlier in the far north-eastern corner adjacent to the River Thames and isolated chalk hills such as Cholsey Hill and the Sinodun Hills, the distinctive scarp drops steeply down to the Vale of White Horse. The plain, a fairly flat ledge at the foot of the high downs linked to a steep escarpment, is characterized by a lack of surface water, with few settlements and large arable fields. Along the lower part of the scarp slope, at the boundary between the porous chalk above and the lower impervious clay, there are springs and a patchwork of woodland and pasture, including significant areas of remnant chalk grassland (such grassland is also found in open downland and to a lesser extent in downland with woodland). Here, there is a series of spring-line villages including Liddington, Hinton Parva, Bishopstone, Ashbury and Letcombe Bassett. On the steep scarp slopes there are small linear ridges or terracettes, sometimes known as 'sheep tracks',

▲ Picturesque thatched cottages in Ashbury on the northern limits of the North Wessex Downs AONB

GEOLOGY AND THE LANDSCAPE

which have been formed by downward soil-creep over millennia.

This is an ancient landscape used by humans for thousands of years. The Ridgeway travels along much of the scarp top, linking many archaeological sites including Neolithic barrows, Iron Age hill forts and the unmistakable outline of the Uffington White Horse. The plain also includes the large flat area of Lower Chalk around Avebury in the west of the AONB, which was extensively used in prehistory and is home to a number of famed archaeological sites.

Much of the area was arable before a change to pastoral use generated the classic sheep-grazed downland that is so characteristic of the region – although arable farming returned in the post-war years owing to modern ploughing and the use of fertilizers. However,

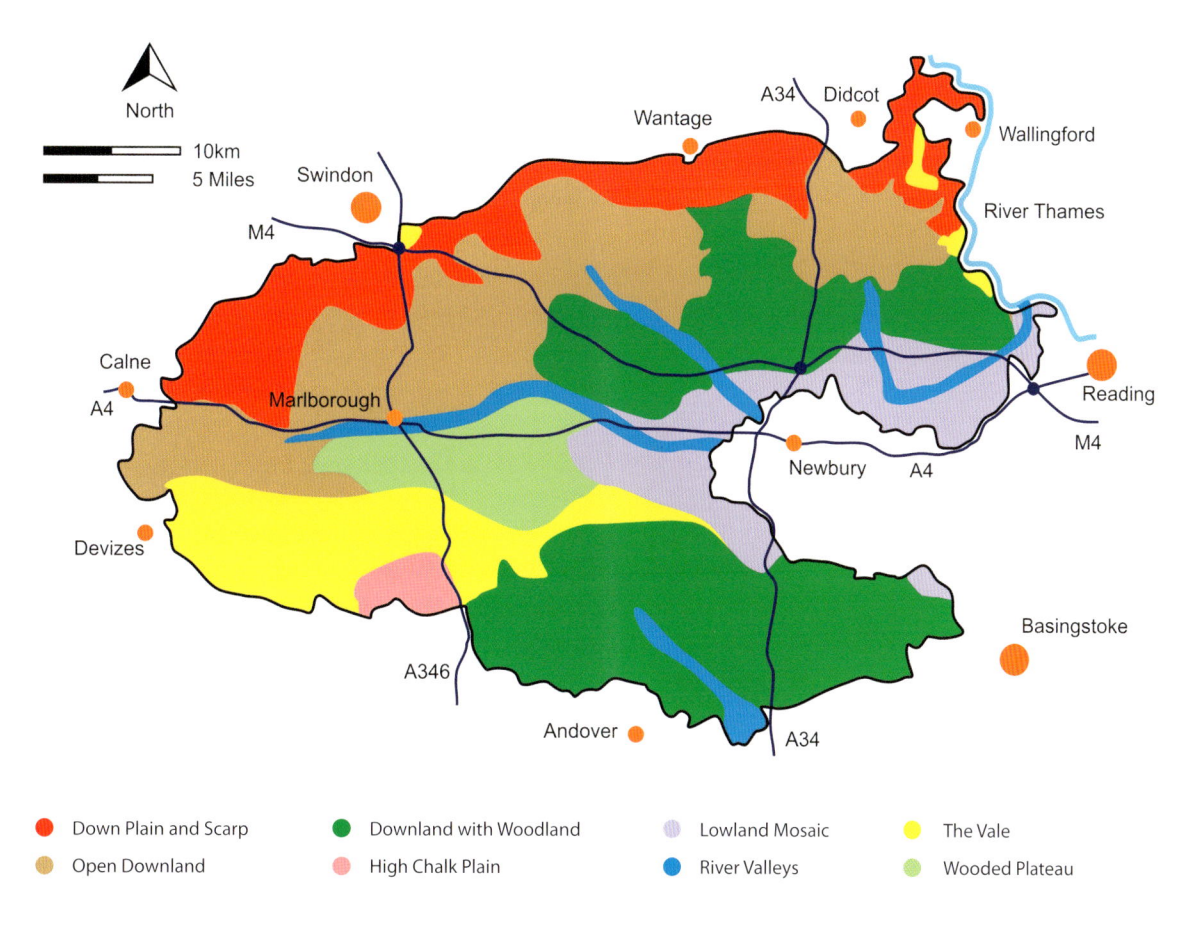

▲ Bishopstone, one of the spring-line villages that lie below the steep northern scarp

Legend:
- 🔴 Down Plain and Scarp
- 🟤 Open Downland
- 🟢 Downland with Woodland
- 🟣 High Chalk Plain
- 🟪 Lowland Mosaic
- 🔵 River Valleys
- 🟡 The Vale
- 🟩 Wooded Plateau

▲ Landscapes of the North Wessex Downs

19

▲ Large open arable fields near Letcombe Bassett

the steep scarps were unsuited to ploughing and these remain as nationally important areas of open chalk grassland, rich in flora and home to numerous species of butterfly. Yet in some areas strip lynchets can be seen; these man-made terraces were formed to allow cultivation to be extended up the hillsides, thereby enabling more land to be ploughed.

OPEN DOWNLAND

Moving south from the distinctive scarp is the remote core of the North Wessex Downs consisting of open downland on the hard Middle and Upper Chalk layers, which form an elevated plateau of open, rounded hills dissected by dry valleys.

This is an area with little surface water and limited habitation, but a high number of racehorse training stables that make use of the springy turf that covers parts of the area. The main land use is arable farming, with very limited tree cover save for the occasional linear shelter belt or the distinctive beech clumps crowning hill summits. The steeper slopes are home to remnant areas of chalk grassland.

The area includes the Horton Downs, Marlborough Downs, Lambourn Downs and Blewbury Downs along with Lardon Chase (a Site of Special Scientific Interest – SSSI), which represents one of the largest remaining fragments of unimproved chalk grassland on the Berkshire Downs. This area also includes the distinctive sarsen boulder fields such as Fyfield Down.

Prehistoric man has left his mark, with hill forts at Oldbury, Rybury and Martinsell Hill, long barrows at West Kennet and Adam's Grave as well as the slightly more recent Saxon Wansdyke.

DOWNLAND WITH WOODLAND

This lies further to the east in two distinct areas separated by the Kennet Valley; the area around Brightwalton and Ashampstead to the north and the Hampshire Downs to the south. A distinctive escarpment, taking in Watership Down and Walbury Hill, defines the northern edge facing towards the Kennet Valley with the bed of Upper Chalk dipping away southwards.

Agricultural land use is more varied here, with a

▲ Trees in winter on the North Hampshire Downs near Watership Down

mixture of arable and pasture, and widespread scattered farmsteads and small villages hiding in the sheltered valleys or located on ridge tops.

WOODED PLATEAU

This landscape is centred on the extensive woodland tract of Savernake Forest, bordered to the north by the Kennet Valley and to the south by the Vale of Pewsey; this was once part of the medieval royal hunting forest of Savernake that was established by the time of the Domesday survey and stretched from East Kennett in the west (encompassing the present West Woods) to as far east as Hungerford. Throughout this gently sloping plateau a thick covering of clay-with-flints results in damp and heavy soils.

Being a former royal hunting forest, settlement is limited to the east of the area along the valley of the River Dun, including Great Bedwyn and Little Bedwyn. The forest today consists of large areas of semi-natural ancient woodland with magnificent oak and beech trees, wood pasture with veteran trees, and also eighteenth- and nineteenth-century beech plantations and some modern coniferous planting. The area supports lichen flora, diverse plant communities, nationally scarce butterflies and moths and a diverse range of birds and mammals.

HIGH CHALK PLAIN

Just nudging into the south-west corner of the North Wessex Downs is a small area of high chalk plain, formed from the northern-most tip of Salisbury Plain. This forms an open landscape, devoid of settlement and under arable production with a steep escarpment forming the northern boundary at Pewsey Hill, providing panoramic views across the Vale of Pewsey immediately to the north.

THE VALE

This landscape is characterized by the Vale of Pewsey, a low-lying landform of upper greensand running east–west and separating the two main upland areas of chalk that dominate the North Wessex Downs. The area has a number of streams rising from the junction between the porous chalk and the underlying clays, which force the

▲ The River Kennet near Chilton Foliat

water table to the surface. Rich loamy and alluvial soils create a productive agricultural landscape with a mixture of arable and pasture now replacing a once predominantly pastoral scene important for cattle and dairying. It is often claimed that the stark difference in Wiltshire's geography, namely the high chalk downland and the low-lying lush valleys, gave rise to the saying 'chalk and cheese', the 'chalk' referring to the chalk downland such as the Marlborough Downs, which are home to sheep grazing, and the 'cheese' referring to the fertile pastures along the river valleys like the Vale of Pewsey and other lower-lying areas with their neutral, clay soils, ideal for dairy farming.

This is a well-populated area including the large village of Pewsey and numerous smaller villages such as Bishops Cannings and Alton Barnes. Both the Kennet and Avon Canal and railway from Reading to Westbury pass through the vale, taking advantage of the low-lying ground.

In addition to the Vale of Pewsey, a number of smaller areas of low-lying vale landscape occur, such as at Wanborough below Liddington and along the eastern edge of the North Wessex Downs, taking in parts of the Thames Valley floodplain around Benson, Basildon and Streatley.

LOWLAND MOSAIC

This landscape, concentrated around Newbury, is largely of medieval origin. Here we have a mix of ancient semi-natural woodlands, plantations, remnant heathland and more open farmland areas, dissected by the Rivers Kennet, Lambourn and Pang and underlain by Tertiary deposits including sands and gravels of Reading and Bagshot Formations. The network of ancient semi-natural woodland, hedgerows and areas of parkland, including wood pasture (areas of open woodland, offering grazing for animals) and veteran trees (trees that are large and/or old), create considerable ecological interest. This is a much more intimate landscape than the wide open spaces of the downs and is one of the most densely populated landscapes within the North Wessex Downs. Of particular interest are the former medieval deer parks such as Englefield, Highclere and Hampstead, many of which were re-fashioned in the eighteenth century as formally designed landscapes.

RIVER VALLEYS

The final land type of the North Wessex Downs consists of the river valleys: although all the valleys drain a much larger area, the narrow corridor immediately beside the rivers has a very distinctive character. These form rather linear landscapes that follow the chalk streams and rivers with a mixture of grazed pastures, water meadows, wetland and woodland. The spring-fed streams and rivers that meander through the downs are rich in plant and animal communities, due to the quality of the mineral-rich, alkaline water that is both naturally clear and fast flowing.

The four main river valleys are the Bourne, Kennet, Lambourn and Pang, the first-named being the exception as it drains south to the River Test, while the other three drain eastwards to the River Thames.

Historically, the main settlements of the region were concentrated in these river valleys as these were where the only form of accessible water in an otherwise dry downland landscape was located. Water meadows were carefully constructed with a network of channels and drains to cover the surface of the meadow with a shallow, relatively fast-moving, sheet of water during the winter. This protected the grass from frost and stimulated early growth, providing feed for sheep. Other distinctive features relating to past management include watercress beds – some of which are still operational – and remains of mill systems; in the next section we look more closely at these chalk streams and rivers.

Chalk Streams and Rivers

The 'gin-clear' waters of our chalk streams are an iconic feature of the English landscape and form a globally rare habitat; fortunately the North Wessex Downs has some excellent examples.

All chalk streams form in the same way, with rainwater falling on the downs; this slowly filters down through the permeable layer of chalk to re-emerge as springs. In the North Wessex Downs around 28–32

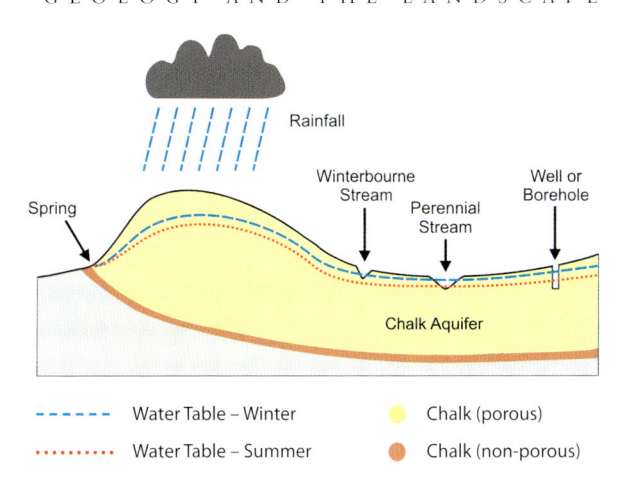

▲ The chalk aquifer and chalk streams

inches of rain falls on the downs and although much of this evaporates or is taken up by plants, around 40 per cent seeps down into the chalk through a network of pores and fine cracks until it reaches a level known as the water table; this is produced by an impervious layer of clay further under the chalk. Below the water table the pores in the chalk, which act like a giant sponge, are saturated with water and form the chalk aquifer – think of it as a natural form of giant underground water storage system. The rate at which this rainwater filters through the chalk is typically 3 feet or so per year and it may take decades for rainwater seeping into the chalk to reach a spring or stream; rates can be much higher when the rock is more fractured.

At the base of the chalk escarpment and in the valleys, the water table reaches the surface to form springs. The height of the water table varies throughout the year, falling during the drier months of the summer and autumn before rising again in response to rainfall in the winter and spring months; in a well at Rockley in Wiltshire, to the north-west of Marlborough, detailed records from 1932 to the present day show that the water table varies throughout the year from around 7 to 60 feet below the ground surface. This seasonal variation means that the upper reaches of many streams only flow during the wetter months; these are known as winterbournes and many of the streams in the North Wessex Downs, such as the River Lambourn from its source to Great Shefford, exhibit this behaviour.

Low winter rainfall can lead to the water table being lowered to such an extent that rivers dry up, whereas, if the water table rises too much, it can rise above the level of the valley floor and cause sustained groundwater flooding.

Chalk streams offer a slightly alkaline water supply with a stable temperature and high clarity owing to filtration through the chalk; because of this they sustain a unique ecosystem. Mid-channel plants are dominated by flowing vegetation such as water-crowfoot (a member of the *Ranunculus* family of plants which also includes buttercups) which has long trailing green foliage and white flowers that form snowy carpets across streams in summer. While along the margins there are lesser water-parsnip and watercress.

Chalk streams support a range of freshwater fish that humans have exploited throughout history. However, it was not until the nineteenth century that these streams and rivers became noted for their brown trout fishing using a form of fly-fishing known as 'dry-fly'; this imitates a real insect or larva that the fish feed on. The abundance of trout is directly linked to the quality and quantity of water in the river.

Nowadays, however, many of these streams and rivers are threatened by ground water abstraction to supply water for public and industrial water supplies, which can drastically reduce the flow of water. This, combined with phosphate and nitrate pollution from sewage treatment works, farm animals and crop production, can have a serious impact on the sensitive ecosystem of a chalk stream.

Along the northern edge of the North Wessex Downs, below the steep north-facing scarp slope, is a line of springs that have formed at the junction of the porous chalk and the lower impervious layers of clay. These springs generate streams such as the Letcombe Brook, which flow northwards into the Vale of White Horse to join the River Ock, which then flows eastwards to join the River Thames at Abingdon.

The 45-mile-long River Kennet, which rises at several locations including Swallowhead Springs just to the south of Silbury Hill near Avebury, flows eastwards to join the River Thames at Reading (other sources include Uffcott about 5 miles to the north of Avebury and near Yatesbury to the north-west of Avebury). The Kennet, which is the most important river of the North Wessex Downs, is joined by several other chalk rivers including the following:

The River Og rises near the hamlet of Draycot Foliat and flows south through Ogbourne St George, Ogbourne St Andrew and Ogbourne Maizey to Marlborough, where it joins the River Kennet.

The River Lambourn rises near Lambourn and flows south-east along the Lambourn Valley, the 'winterbourne' section passing through Eastbury and East Garston before the river becomes perennial near Great Shefford, from where it flows south-east through Welford and Boxford to join the River Kennet at Newbury. The cricket commentator Howard Marshall (1900–1973) wrote about fly-fishing on the River Lambourn in his book *Reflections on a River* (1967).

▲ The River Lambourn at Maidencourt Farm between East Garston and Great Shefford (the river around here is known as a winterbourne)

GEOLOGY AND THE LANDSCAPE

▲ The River Pang at Bradfield

The River Dun starts out near Great Bedwyn in Wiltshire and flows north-east into Berkshire, joining the River Kennet at Hungerford.

The River Enborne rises near the villages of Inkpen and West Woodhay, to the west of Newbury, and joins the River Kennet near Aldermaston; part of its course forms the border between Berkshire and Hampshire. The Enborne has a literary claim to fame as this is the river that a group of rabbits have to cross in Richard Adams's classic novel *Watership Down*.

The upper River Kennet, from its source down to Woolhampton, forms a wonderful example of a chalk stream and because of this it has been designated a Site of Special Scientific Interest (SSSI) for its range of rare plants and animals that are unique to chalk streams. The lower reaches, which are navigable by boat, are known as the Kennet Navigation, which forms part of the Kennet and Avon Canal.

Unfortunately the Kennet suffers from a high level of water abstraction – at the time of writing, Thames Water were permitted to abstract an average of up to 11.1 mld (million litres per day) from the Kennet's subterranean aquifer under Axford, located between Marlborough and Hungerford (that is equivalent to 2.9 million gallons, or more than four Olympic-sized swimming pools per day) – water that would otherwise maintain the flow of the river. Once removed, much of the water is pumped outside the Kennet catchment, so it never flows back into the river. In dry summers, this loss of water causes severe stress to the ecology of the river and in some years sections of the river can dry up completely with disastrous consequences for the fish and wildlife.

However, all is not lost: new treatment plants, along with more environmentally aware farming practices, have reduced the amount of pollutants entering the river and both the Environment Agency and Thames Water are looking at ways of reducing the amount of water that can be abstracted. Over time these actions should help to ensure that the River Kennet remains one of the great chalk rivers of southern England.

The River Kennet and its tributaries are by no means the only chalk streams in the North Wessex Downs.

The River Pang, which rises near the village of Compton, meanders its way through Berkshire for 14 miles, passing through the picturesque riverside villages of Bucklebury, Stanford Dingley and Bradfield to join the River Thames at Pangbourne. As with most chalk streams, the upper reaches are termed a winterbourne, tending only to flow in the wetter winter months. The perennial head of the river is at Kimberhead Spring and the Blue Pool near Stanford Dingley, where the clear spring water from the artesian aquifer bubbles to the surface throughout the year.

The River Swift rises from a spring at Upton before disappearing back into the chalk further downstream, and then re-emerges at Hurstbourne Tarrant as the Bourne Rivulet (not to be confused with the River Bourne near Burbage that joins the River Avon) – often just called 'The Bourne' by locals. The river then flows south-eastwards through the villages of Hurstbourne Tarrant, St Mary Bourne and Hurstbourne Priors before joining the River Test near Tufton; this is the only tributary of the River Test to start in the North Wessex Downs.

Like the River Enborne, the Bourne Rivulet has its place in literary history, having been described by Harry Plunket Greene in his book *Where the Bright Waters Meet* as 'unquestionably the finest trout stream in the south of England'. His 'bright waters' are the confluence of the Bourne Rivulet and the River Test just to the south of Hurstbourne Priors, and the book takes a snapshot of the early twentieth century, including mention of the growing of watercress on chalk streams: 'The watercress beds above the viaduct have scarred her face and marred her beauty forever'. Watercress is still grown commercially along the river, though more environmental awareness these days has led to the reinstatement of the stream to help maintain the ecology of the river.

The River Avon, often known as the 'Hampshire' or 'Salisbury' Avon to distinguish it from the other River Avons in England, begins as two separate rivers in Wiltshire: the Western Avon, which rises to the east of Devizes, and the Eastern Avon, which rises just east of Pewsey. These two rivers merge at Upavon before continuing southwards to the English Channel. One tributary of the River Avon is the River Bourne, which rises at the eastern end of the Vale of Pewsey, close to the village of Burbage. The river flows south, cutting through the chalk escarpment at Collingbourne Kingston and then continues across Salisbury Plain to join the River Avon at Salisbury.

CHAPTER TWO

History – from the Stone Age to the present day

A Potted History

EARLY BEGINNINGS

Our journey through the history of the North Wessex Downs starts way back in the Palaeolithic or Old Stone Age period. Around 12,000 years ago the last Ice Age was coming to an end, the climate was warming up, trees were spreading northwards and sea levels were starting to rise.

Britain was still connected to Continental Europe by a flat open plain that had been formed by a giant 'super river' caused by water released from a massive glacial lake about 450,000 years ago, situated over what is now the southern North Sea. This river carved its way through the chalk hills that once ran in a continuous line connecting southern England to northern France; a second massive flood around 180,000 years ago caused further erosion at the Strait of Dover. During interglacial periods (the warmer periods between ice ages) this channel that had been carved out by these massive rivers was flooded by rising sea levels caused by the melting ice sheets, making Britain an island. However, before the sea levels rose for the final time this land-bridge allowed animals and people to migrate from Europe to Britain: these were our hunter-gatherer ancestors, living in caves and temporary camps and using simple stone spears and knives.

The Mesolithic period, from 9000 BC to 4200 BC, saw the hunter-gatherer lifestyle flourish, although towards the end of this period both pottery and farming had started to make an appearance. It was towards the start of this period that the rising sea levels flooded the plain

▲ Knap Hill, site of a Neolithic causewayed camp (see page 45) and the southern edge of the Pewsey Downs looking out over the Vale of Pewsey

linking Britain and Europe for the last time to form what is now the English Channel.

Next came the Neolithic period from 4200 BC to 2200 BC, which witnessed further development of the farming lifestyle, with more permanent camps being constructed, along with massive henges and long barrows. This is the period that left the North Wessex Downs with many important monuments such as the Early Neolithic causewayed enclosures at Windmill Hill, Knap Hill and Rybury and the Neolithic long barrows such as the West Kennet Long Barrow and Wayland's Smithy.

The Late Neolithic and Early Bronze Age (around 2400 BC) brought about the construction of some of the iconic prehistoric monuments of southern England, including further causewayed enclosures, burial mounds, the great henge at Avebury and the remarkable Silbury Hill. It was during this period that the distinctive grooved ware pottery came into use. Farming continued to develop and the characteristic open 'chalk downland' vegetation started to appear over parts of the North Wessex Downs as more land was used for grazing livestock.

THE DISCOVERY OF BRONZE AND IRON

The Bronze Age (2200–750 BC) was brought about by the discovery that bronze could be made by mixing a small amount of tin (10 per cent) with copper (90 per cent). Bronze is much harder than copper and over the next few centuries it gradually replaced stone as the main material for both tools and weapons.

The Early Bronze Age was the time of the Beaker people, named after their distinctive inverted bell-shaped clay pots, who came over from mainland Europe and may well have brought with them the knowledge of making and working with bronze.

Beaker people mixed with the native Neolithic farmers, and beakers have been found in many of the ancient burial sites within the region. They were involved in the further development of the Avebury henge and Silbury Hill and gave rise to what is termed the Wessex Culture, a name given to rich grave goods found in round barrows in southern Britain, including precious ornaments of amber and gold. It was during the Early Bronze Age that many of the round barrows, which are such a characteristic feature of the North Wessex Downs, were constructed.

Iron-working techniques reached Britain from Southern Europe, heralding the start of the Iron Age (750 BC to the Roman Invasion of AD 43). Iron was stronger than bronze and it revolutionized many aspects of life, none more so than in agriculture where iron-tipped ploughs could dig the land more easily than old wooden or bronze ones, and iron axes could fell wood more efficiently. The landscape changed to one of arable farming, pasture and managed woodland. Enclosed settlements, or hill forts, were built in prominent locations and the remains of these Iron Age hill forts can be seen throughout the North Wessex Downs.

During this time, sheep were being grazed on open downlands that were now largely devoid of woodland. Reared for both meat and wool, sheep require less water than cattle and were suited to life on the downs, while cattle may have been grazed in valleys close to water sources. Fields were cultivated with crops including wheat, barley, oats and rye.

THE ROMAN INVASION

Although the armies of Julius Caesar made small-scale invasions of Britain in 55 and 54 BC, it was not until AD 43 that the Roman Conquest took hold. Following this a distinctive Romano-British culture emerged which, over the next few decades, spread as far north as Hadrian's Wall.

Despite the Roman rule lasting for nearly four centuries, little obviously visible evidence of their occupation remains within the North Wessex Downs. They did, however, lay down a network of roads linking their main towns so that soldiers and supplies could be moved with relative ease, and several of these roads passed through the North Wessex Downs. A quick look at an Ordnance Survey map shows not only the course of these former roads but also highlights the fact that some of our current roads follow routes laid down by the Romans.

However, we know from aerial photography and excavations that the Romans left more than just roads. A number of villa sites have been identified within the North Wessex Downs, with undoubtedly the best-known being at Littlecote House, where the Orpheus Mosaic was found. Also, the former fortified settlement of *Cunetio* just to the east of Marlborough is noted for the largest find of Roman coins in Britain.

THE MEDIEVAL YEARS

The demise of the Roman Empire in Britain, around AD 410, paved the way for the Anglo-Saxon period, which lasted until the Norman Conquest of England in 1066. During this time, settlement migrated from the chalk downs to the river valleys, exploiting the abundant water supplies. Agriculture shifted, with sheep farming becoming more dominant on the open downs and cultivation of the heavier clay soils developing in the low-lying areas.

It was towards the start of this period, in the late fifth century, that the massive earthwork known as the Wansdyke was constructed, stretching west across the Marlborough Downs above the Vale of Pewsey from Marlborough to Morgan's Hill. The earthwork probably gained its name from the Saxon god Woden, hence it became 'Woden's Dyke' and, over time, Wansdyke.

Liddington Castle is believed by some to be the site of the Battle of Mons Badonicus (or Mount Badon) where the Britons, reputed by some to have been led by the legendary King Arthur, defeated the invading Saxons sometime around AD 500. However, in AD 556, the tables were turned at the Battle of Beranburgh (Beran Byrig) when the Saxons, led by Cynric and his son Ceawlin (who later became King of Wessex in AD 560), were victorious; the site of the battle is claimed to be Barbury Castle, a few miles south of Swindon.

The Saxon invasions led to changes in the Romano-British culture and by the end of the sixth century England had become divided into several kingdoms, namely Northumbria, Mercia, East Anglia, Essex, Kent and Wessex.

These early Saxon settlers were pagan – worshipping several gods – and Christianity, which had started to flourish in Romano-British culture, was driven to the more western parts of Britain such as Wales and Cornwall where it continued to thrive. In AD 597 Pope Gregory sent a mission to England to bring Christianity to the Anglo-Saxons and during the seventh century large parts of the country converted. Scholarly monks taught Latin and new churches and monasteries were built.

It is during this time that the Saxon royal hunting forests of Savernake, Chute, Freemantle and Pamber were formed. These were not like a modern forest full of trees, but rather a mosaic of woodland, scrub and heath.

During the ninth century, Danes were invading parts of England and in AD 871 Alfred the Great (849–899), who was born at Wantage, defeated the Danes at the Battle of Ashdown in Berkshire ('Ashdown' was the

▲ Statue of Alfred the Great at the Market Place in Wantage (see page 102)

ancient name for the whole expanse of the Berkshire Downs); later that year he succeeded his brother, Ethelred, as King of Wessex. Despite his success at Ashdown, the Danes continued to attack Wessex, forcing Alfred further west. Following his victory at the Battle of Ethandun (Edington) in AD 878, Alfred negotiated the treaty of Wedmore with Guthrum and the Danes; Guthrum converted to Christianity and many of the Danes returned to East Anglia. In AD 886 Alfred negotiated a partition treaty with the Danes dividing England, with part becoming Danish territory – known as the 'Danelaw' – and Alfred controlling Wessex, West Mercia and Kent.

In about AD 890 Alfred built up the defences of his kingdom and constructed a series of well-defended settlements across southern England to ensure that it was not threatened by the Danes again. He commissioned the Anglo-Saxon Chronicle (this was maintained and added to until the middle of the twelfth century), funded church schools, brought in a code of laws and developed his capital at Winchester, where he is buried. Alfred's successor, Edward the Elder, then became overlord of England.

By the time of Ethelred the Unready, the Danes were still sending raiding parties and, in 1015, Cnut (Canute) arrived with a strong army to take control of London, Wessex, Mercia and Northumbria. He became king in 1016, but on his death in 1035 feuding between his sons allowed Godwin of Wessex to acquire the position of kingmaker and secure the throne for Edward the Confessor.

On Edward's death Harold, Godwin's son, laid claim to the throne of England. Having successfully defeated a Viking invasion in the north of the country (at Stamford Bridge) he marched south again and was defeated by the invading Normans at the Battle of Hastings in 1066. By late October Wallingford had fallen to the Normans and two months later William the Conqueror was crowned King of England, heralding the Norman period.

The Domesday survey of 1086 provides an insight into how the English landscape was organized both prior to and following the Norman Conquest, which was a pivotal event in English history, largely removing the native ruling class and replacing it with a foreign aristocracy loyal to William. These aristocrats established motte and bailey castles throughout England during the eleventh century to act as local centres of power and control. These consisted of a large earthen mound topped with a wooden stockade, known as the 'keep', surrounded by a ditch or fortified enclosure, known as the bailey. These early castle mounds can be found at several places in North Wessex (see Medieval Castles and Civil War on page 49).

However, the Normans did more than just build castles; along with the Crown, the Church exercised great power and numerous monasteries and churches were built, characterized by Romanesque rounded arches over windows and doorways. Many of the present churches within the North Wessex Downs originate from the eleventh and twelfth centuries, albeit with a fair number of alterations over the centuries.

Prosperity and growth in the late twelfth and thirteenth centuries led to the rapid expansion of towns both within and just outside the AONB, yet by the early fourteenth century harvest failures and famines gave way to the Black Death around 1350 – one of the most devastating pandemics in human history. The Black Death, or bubonic plague, killed close to half of the population of England, altering the balance of economic and social power between peasants and lords, with serfdom largely disappearing.

POST-MEDIEVAL TO THE PRESENT DAY

By the sixteenth century there was a distinct 'middle class' emerging in the countryside, one contributory factor being that the woollen cloth industry underwent rapid growth. East Ilsley was famed for its sheep fairs; Newbury had a famed clothier, one John Winchcombe (d.1519), aka Jack of Newbury; and many fulling mills were built alongside the rivers that provided the power to run them.

In 1534, Henry VIII broke with Rome, establishing the Church of England and bringing about the Reformation so that he could marry Anne Boleyn. The following years saw the Dissolution of the Monasteries, which not only transferred the ownership of large areas of land to the Crown, such as Pangbourne Manor, but also into private hands such as the Manor of Beenham passing to Sir Henry Norreys. Both of these manors had, until then, been held by Reading Abbey.

With the seventeenth century came the English Civil War and a series of armed conflicts between Parliamentarians and Royalists that ultimately led to the trial and execution of Charles I. Castles that had been Royalist strongholds, such as Donnington and Devizes, were later destroyed by order of Parliament.

The eighteenth century witnessed changes including the widescale introduction of farm machinery. Jethro Tull invented the seed drill in 1701 (see page 60) and late in the century threshing machines were invented. These drastically reduced the number of farm labourers required. These advances in farm equipment, especially

▲ The polissoir stone at Fyfield Down was used by Neolithic people to smooth stone axes

the horse-powered threshing machine, coupled with the failure of crop harvests, led to the 'Machine Riots' in 1830, with labourers setting out to destroy farm machinery in an attempt to get fairer living standards. However, during this period farmland was also improved through drainage and the use of fertilizer.

Lace and silk-making developed in places such as Marlborough and Whitchurch, but during the late eighteenth and nineteenth century the cloth industry declined markedly.

Throughout the early nineteenth century the countryside suffered depopulation while town populations grew rapidly and, by the end of the century, more people were working in industries based in towns than were working in agriculture. Further declines in the rural population followed the First and Second World Wars, with increased mechanization on farms.

The transport network in the North Wessex Downs area had started to improve from the late seventeenth century with the introduction of Turnpike Trusts and, in 1810, canal mania arrived with the opening of the Kennet and Avon Canal, linking Reading and Bath. This was quickly followed by the arrival of the railways that served as major trade and communication routes. Following the end of the Second World War the number of road vehicles steadily increased, the road network continued to improve, including the upgrading of the main north–south route – the A34 – and the opening of the M4 motorway, which slices east to west through the region, in the early 1970s.

A further significant event in the history of the North Wessex Downs was its designation as an Area of Outstanding Natural Beauty in 1972, with the aim of conserving and enhancing the natural beauty of the downs.

Archaeology Uncovered – from Causewayed Camps to Medieval Castles

The North Wessex Downs is an ancient landscape, littered with monuments built by our prehistoric ancestors. The Neolithic and Bronze Age periods left behind some of the most impressive sights: the great henge and

Silbury Hill at Avebury, magical Wayland's Smithy and the amazing galloping outline of the Uffington White Horse looking out over the Vale of White Horse, while at Fyfield Down NNR there is a sarsen stone, known as a polissoir, that was grooved and smoothed over hundreds of years by our Neolithic ancestors as they polished stone axes.

From the Iron Age we have the remains of numerous hill forts that consist of concentric rings of earth ramparts and ditches, some having only a single ditch and rampart, others having two, and some boasting three defensive rings; due to their prominent location, many of these forts offer the visitor some stunning views across the region. The Romans, too, left a legacy of ancient roads and hidden treasures while the early Anglo-Saxons gave us impressive territorial boundaries such as the Wansdyke.

To learn more about our prehistoric ancestors, one can visit the Wiltshire Heritage Museum in Devizes. This museum houses an important collection of Neolithic and Bronze Age artefacts, along with a fine collection of Iron Age material that includes the Marlborough Bucket, found in a grave in Marlborough in the early nineteenth century.

Unfortunately, a fair number of ancient sites have been 'lost' to ploughing in the twentieth century. These now exist only as crop markings that can be seen in aerial photographs: such a site is the Neolithic causewayed camp on Rams Hill close to the Ridgeway. Yet, despite this destruction, there are still plenty of truly impressive sites to be explored.

AVEBURY WORLD HERITAGE SITE

At Avebury – lying in the heart of the downs to the west of Marlborough – is a collection of Neolithic and Bronze Age sites which, combined with Stonehenge, form a UNESCO World Heritage Site – an area containing the largest number of surviving Neolithic and Bronze Age monuments in Western Europe. These are the jewels in the North Wessex Downs' crown and they form one of the most impressive prehistoric sites in Britain – maybe this was a 'sacred landscape', whose purpose is still difficult to comprehend.

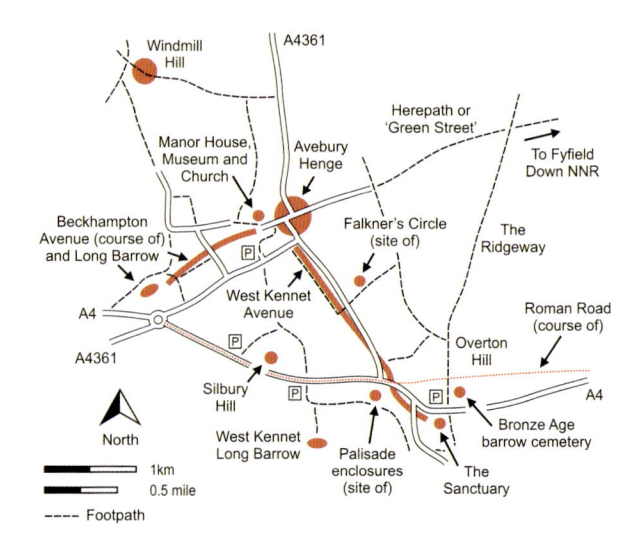

▲ Map of the Avebury World Heritage Site

Within the Avebury World Heritage Site we have Windmill Hill, the Avebury henge and stone circle, West Kennet Avenue, the Beckhampton Avenue, The Sanctuary, mysterious Silbury Hill, West Kennet Long Barrow and Overton Hill Bronze Age barrow cemetery.

Let's start our tour through the landscape of Avebury at Windmill Hill (SU086713), a classic example of a Neolithic 'causewayed enclosure' that was first occupied around 3800 BC. The three concentric rings of ditches circling the hilltop, enclosing 21 acres, are all intermittent, being broken at intervals by crossing points, hence the term 'causewayed'. The hill was later used as a Bronze Age cemetery.

The site was first recorded in the eighteenth century and a local clergyman, the Revd H. G. O. Kendall, undertook the first excavations in 1922–3, which brought Avebury to the attention of one man in particular – Alexander Keiller, heir to the Dundee Marmalade company. Keiller bought most of the site in the 1920s, saving it from being used as a wireless relay station, and undertook a series of detailed excavations. Large quantities of animal bones were found, along with numerous flint tools and fragments of pottery; some of these objects can be seen in the Alexander Keiller Museum at Avebury. Through these excavations and finds we now know more about Windmill Hill and its role in Neolithic life. The camp, probably not unlike other causewayed camps, was used as a meeting place, being occupied only at certain times rather than continuously. People would travel from small outlying settlements to trade goods, anything from clay pots to stone axes and livestock, and it is not difficult to visualize important feasts and maybe ceremonies taking place here; certainly the quantity of animal bones suggests that a large number of animals were killed here for both meat and skins.

Construction of the Avebury henge (SU101699), one of the largest and most complex Neolithic henge monuments in Britain, began around 2800 BC and spanned the next 600 years.

The henge consists of a huge circular outer bank and inner ditch, which now includes part of Avebury village. Within this henge is a circle of standing stones, enclosing two smaller stone circles, each with a central feature. Although it looks impressive today, when originally constructed it must have been a truly awe-inspiring sight – time has mellowed its appearance. The present ditch is 10–13 feet deep, yet excavations in 1914 revealed that the ditch was originally some 30 feet deep, with much steeper sides and a 13-foot-high bank.

Within the henge a later outer stone circle consisting of around a hundred standing stones was built, along with two smaller inner circles which had around thirty stones in each; within each of these smaller circles were more standing stones.

Why it was built at this site, and for what purpose, remain mysteries: many suggest it was built for observing the heavens, the changing seasons or the cyclic variations in the moon and sun, yet, from the evidence we have today, it is impossible to be sure.

From the writings of William Stukeley (1687–1765) in the eighteenth century we know that he saw first-hand the destruction of many of the stones that made up the Avebury stone circle – to be used as a local source of building stone by lighting large fires around them and then dousing them in cold water, thus cracking the stone into smaller blocks. He writes of that destruction: 'And this stupendous fabric, which for some thousands of years, had brav'd the continual assaults of weather, and by the nature of it, when left to itself, like the pyramids

▲ One of the large sarsen stones that forms the stone circle at Avebury

of Egypt, would have lasted as long as the globe, hath fallen a sacrifice to the wretched ignorance and avarice of a little village unluckily plac'd within it.'

Stukeley, who studied medicine at Cambridge University, had a keen lifelong interest in making architectural drawings and sketches of historical artefacts and he continued with this alongside his career as a doctor. From his travels around Britain he published *Itinerarium Curiosum* in 1724 and his works about Stonehenge and Avebury appeared in 1740 and 1743. Having witnessed the destruction of the stones, Stukeley knew many of the 'stone-smashers', none more so than local farmer Tom Robinson, whom he christened 'The Herostratus of Avebury'.

We also know from the earlier works of John Aubrey (1626–1697), the man who 'discovered' Avebury in 1649, that many of the stones were still standing at that time – his plan shows close to seventy stones in situ and many more lining the long avenues towards Beckhampton and The Sanctuary.

Just imagine what the stone circle would have looked like today if it had been left untouched – unfortunately the clock can never be turned back fully. Yet, despite the destruction, the appearance of the henge as we see it today is mainly down to the efforts of one man, Alexander Keiller, who bought and saved the site in the late 1930s.

At the start of the twentieth century only fifteen stones were left standing, yet through his excavations, Keiller was able to locate and re-erect a further twenty-one stones. No one really knows why many of the stones were buried in medieval times; maybe they were buried so that the land could be used for agriculture, or as an attempt to expunge pagan connections with the site, perhaps in the belief that it was the work of the Devil himself. In the case of one unfortunate soul, the very standing stone that he was trying to topple (which probably weighs around 13 tons) fell on top of him, crushing him in the hole that had already been dug for the stone … or maybe foul play was at hand. Keiller's excavations found the body along with several silver coins dated from the early fourteenth century, a pair of scissors and a surgical lance or probe; these latter two objects suggested to some that he was a travelling barber-surgeon. Many of Keiller's finds are displayed in the museum he opened in an old stable building beside Avebury Manor where he once lived.

In the north-east quadrant of the henge, close to the road, are two of the tallest stones at Avebury. These stones formed part of a cove – this being the term for a tight grouping of standing stones – and again, through the work of William Stukeley, we know that until 1713 there used to be three stones; the third stone (which was destroyed) standing as a pair to the tall, slender southern stone.

To the south-east of the henge is The Sanctuary (SU118680). Construction began sometime between 3000 and 2500 BC of what was originally a complex circular arrangement of timber posts which may have stood to some height, with the later addition of a double stone circle over 130 feet in diameter. Its function, like so many prehistoric sites, remains a mystery, although many human bones, accompanied by food remains, have led some to speculate that this was once the home of a revered person where elaborate death rites and ceremonies may have been performed.

Again, from the writings of John Aubrey in 1649, we know that many of the stones were still present in his day, yet by the time of William Stukeley, some seventy years later, the destruction of the site was already underway. Indeed the destruction was so great that the site was completely destroyed and lost in the mists of time, only being rediscovered in the 1930s using Stukeley's original illustrations. All that remains today are concrete blocks marking where the timber posts and sarsen stones once stood.

Linking the henge and The Sanctuary is the 1½-mile-long West Kennet Avenue, built shortly after the stone circles were added to The Sanctuary. Running in a south-easterly direction, close to the present B4003, this avenue originally consisted of around a hundred pairs

▲ Looking north-west along West Kennet Avenue

▲ West Kennet Long Barrow

of standing stones, standing about 50 feet apart. As with many of the ancient sites in the area, many of the stones were either buried or destroyed.

Partway along the West Kennet Avenue, a couple of hundred yards or so to the east of the road, is Falkner's Circle (SU109693). The circle, named after the man who first identified the site in 1840, once consisted of twelve stones making a 118-foot-diameter circle; sadly only one stone now remains. The site was partially excavated in 2002 and evidence of burning pits and stone destruction were found.

The West Kennet Avenue was not the only standing stone avenue at Avebury. Leaving the henge along the present-day High Street was another avenue of paired stones – The Beckhampton Avenue – running in a south-westerly direction towards the present-day Beckhampton. Only a few stones had survived by the eighteenth century and many doubted its existence. However, recent testing has confirmed that at least four further stones have been buried and evidence was found confirming where five more stood before they were destroyed. Lying at the end of the avenue are two standing stones known as the Longstones. One is called 'Eve' and formed part of the avenue itself, while the other – 'Adam' – formed part of a cove. Close by is the Beckhampton Long Barrow, sometimes called the Longstones Long Barrow, said to be one of the oldest Neolithic long barrows in England.

Just across the A4 from The Sanctuary is Overton Hill and the well-preserved Early Bronze Age (about 2000 BC) barrow cemetery known to the Saxons as *Seofan Beorgas* or Seven Barrows (SU119681). The cemetery, once crossed by the Roman road from London to Bath, contains both bell- and bowl-shaped barrows and excavations of one of the barrows revealed evidence of twelve people; six cremations and six inhumations.

To the south-west of here, up on the skyline, is the West Kennet Long Barrow (SU104677), one of the largest and most impressive Neolithic chambered tombs in Britain. Built around 3600 BC, the east–west aligned earth mound, which tapers away from its chambered

▲ Looking inside West Kennet Long Barrow

eastern end, is slightly over 328 feet in length and up to 10 feet high; the eastern end, which is some 82 feet wide, is faced with large, upright sarsen boulders.

Excavations of the barrow in 1859 and then again in 1955–6 revealed five burial chambers opening off a central passage; two on either side of the passage and one at the end. These chambers contained the remains of over forty people, both male and female, of various ages, placed here over a relatively short period of time. The chambers were later closed, the passageway infilled and three massive sarsen blocking-stones were placed across the east-facing front for the tomb.

To the south-east from here is the tree-covered East Kennet Long Barrow, the largest Neolithic long barrow in Britain (no public access), while, if you look to the north-west, you can see the last, and probably the most mystical, stop on our tour of Avebury's treasures.

Silbury Hill (SU100684), the largest man-made prehistoric mound in Europe at over 130 feet in height and with a base diameter of 548 feet, was built around 2400 BC. It took somewhere between 100 and 400 years to make, using nothing more sophisticated than deer antlers as picks and oxen shoulder blades as shovels. It must have been an important site, yet why our ancestors put so much effort into its construction remains a complete mystery – although some speculate that its construction had something to do with the fact that the River Kennet rises from springs thereabouts. One thing that the experts seem to agree on is that it was not built as some giant burial mound, as apparently it contains no burial.

Over the years several attempts have been made to try to learn more about the hill's construction or, more importantly for some, to find buried treasure. In 1776 Hugh Percy, Duke of Northumberland, financed the digging of a vertical shaft into the hill, while in 1849 the Revd John Merewether and other members of the Royal Archaeological Institute oversaw the excavation of a horizontal tunnel, and Professor Richard Atkinson led a third major investigation between 1968 and 1970.

▲ Mystical Silbury Hill – the largest prehistoric man-made mound in Europe

Unfortunately, none of the tunnels were fully backfilled, which resulted in a large crater appearing on the summit of the Silbury Hill in the summer of 2000. Recent conservation work has been undertaken to stabilize the hill and this has allowed archaeologists one final look inside it, providing new information that may one day lead to a better understanding of why and how it was constructed. (Note that there is no public access onto the hill.)

Over the centuries various legends have been attached to the hill. Some claim it to be the burial place of King Zel (or Sil), a legendary warrior buried in golden armour; some suggest that it could have been some form of solar observatory by means of the shadows cast by the mound on the carefully levelled plain to the north. Another legend suggests that it was made by the Devil himself: either he was going to empty a huge sack of earth on the nearby town of Marlborough, but was forced to drop it at Silbury through the magic of priests at nearby Avebury or, in another version, he was thwarted by a quick-thinking cobbler. However, these are all just stories: no one really knows the truth behind Silbury Hill, which just makes it all the more mysterious.

While on the subject of Silbury Hill, mention must be made of what is now known to be its smaller sister, the Marlborough Mound, about 5 miles to the east (SU183686). Legend claims that this man-made mound is the final resting place of Merlin, the Arthurian wizard. However, recent radiocarbon dating on samples taken from within the mound, located in the private grounds of Marlborough College, have confirmed that it was constructed around 2400 BC, about the same time as Silbury Hill. So now the North Wessex Downs has not one, but two, prehistoric man-made mounds (Marlborough and its mound are mentioned further on page 83).

As a final thought about Avebury (but by no means exclusive to this area), new discoveries are still being made through the careful analysis of aerial photographs

▲ The Devil's Den at Clatford Bottom – the remains of a Neolithic chambered burial chamber

that can 'reveal' hidden features through cropmarks that are no longer visible on the surface of the ground. One such recent discovery was a pair of Late Neolithic palisade (wooden stakes fixed into the ground to form a defensive structure) enclosures at West Kennet (SU110682). Although first photographed in the 1950s, the significance of the cropmarks was not recognized for more than thirty years. Recent excavations have confirmed that the site is over 4,000 years old and fragments of grooved ware pottery were also found. However, as to their purpose, no one is really sure.

OTHER ANCIENT SITES

Let us now leave behind Avebury and take a look at some of the other Neolithic, Bronze and Iron Age sites that can be easily visited within the North Wessex Downs.

Devil's Den

Located to the east of Avebury in Clatford Bottom between Fyfield Down NNR and the A4 are the remains of a Neolithic chambered long barrow known as the Devil's Den (SU152696). From an illustration by William Stukeley in 1723, we know that there were other large sarsen boulders arranged along the sides of the mound that formed the long barrow with the existing Devil's Den at the south-eastern end. These other stones have long since disappeared and the mound has been greatly reduced by ploughing; the stones were re-set in 1921, including a concrete support to stop the structure collapsing.

The form of the Devil's Den – two upright standing stones and a capstone – has led some to suggest that it is the remains of a dolmen or cromlech (a portal tomb that would have been covered in earth) rather than part of a chambered long barrow.

As with many ancient sites, legends have been associated with the Devil's Den, including one about a large black dog with piercing red eyes that guards the site. The great Wessex novelist Thomas Hardy referred to it as the 'Devil's Door' in his short story *What the Shepherd Saw* (1881), as 'a Druidical trilithon, consisting of three oblong stones in the form of a doorway, two on end, and one across as a lintel … locally called the Devil's Door'.

▲ The former Iron Age hill fort at Blewburton Hill

Aldbourne Four Barrows

Located high up on Sugar Hill to the north-west of Aldbourne (SU249773) is a very well-preserved collection of four Bronze Age barrows known as Aldbourne Four Barrows. The barrows, each around 8 feet high, run in a north-west to south-east line; three are known as bell barrows because of their upturned bell-shape and the fourth is a plain bowl barrow, looking like an upturned mixing bowl.

Excavations towards the end of the nineteenth century confirmed that some of the barrows contained cremations while others had skeletal remains. Various goods, including flint arrowheads and beads, were also found; the contents are now on display in the British Museum. Excavations of a nearby bowl barrow unearthed an engraved pottery cup with matching lid – known as the Aldbourne Cup, along with bronze awls, a bronze dagger and assorted beads, which are also in the British Museum.

Membury Camp

Sitting on the border between Wiltshire and Berkshire, just south of Membury Services on the M4, is Membury Camp (SU301753). Situated in the south-western corner of a small plateau, the remaining earthworks are completely shrouded in trees, while the area within the fort is now arable farmland. The camp consists of a single ditch with banks on either side, a gap on the eastern side is most likely to be the original entrance; other gaps appear to be more modern. The east side of the fort has also been partly destroyed by the construction of the Second World War airfield, RAF Membury. Although never excavated, fragments of pottery as well as early Neolithic flint tools have been found in the area.

Grimsbury Castle

Continuing eastwards from Membury Camp, to the south of Hermitage and the M4, is Grimsbury Castle (SU510722). Now surrounded by dense woodland, it is hard to imagine the fort's former strategic position. Excavations in 1957 not only showed how the ditches were originally much deeper but also unearthed various finds including flint tools, 'pot-boilers' (stones heated in a fire and then used to heat water), sling pebbles, a quern stone for grinding grain and fragments of Early Iron Age pottery. Further excavations at the western entrance have shown that it was strengthened and narrowed with the addition of a flint block wall.

Blewburton Hill Fort

Heading north-east brings us to this feature on a low-lying outcrop of the downs just to the east of Blewbury (SU545861). This was originally built as a palisade (timber defensive wall or fence) settlement, one of the earliest types of defensive barrier used on hill forts, though it was later refortified in the sixth and fifth centuries BC. The site consists of several ditches and ramparts on the western side, the eastern half having been affected by ploughing in the past. The fort was probably abandoned before the Romans arrived, although the site was later used as a Saxon cemetery sometime around AD 500.

Wittenham Clumps

Right in the far north-east corner of the North Wessex Downs are the twin tops of the Sinodun Hills, more commonly known as the Wittenham Clumps – Round Hill and Castle Hill, overlooking the River Thames. The south-eastern top, known as Castle Hill (SU569924), is crowned by the remains of an Iron Age hill fort, although the original earthworks date from the Late Bronze Age. The Channel 4 programme *Time Team* investigated the site in 2004 using ground-penetrating radar on the area just south of the tops, and further excavations revealed the remains of a Romano-British villa. Castle Hill was also the location of the Poem Tree – see Writers of Prose and Poetry, page 147.

Grim's Ditch

Running along the northern edge of the Berkshire Downs, heading west from the downs above Aston Upthorpe (SU542830), not far from the River Thames, to Middlehill Down (SU4178412) to the south-east of Wantage, close to the Ridgeway (see below), are fragments of a linear earthwork known as Grim's Ditch. Archaeological evidence suggests that parts of this earthwork were constructed sometime in the Late Bronze Age, probably as a territorial boundary rather than a defensive structure.

The Ridgeway

The Ridgeway, which follows the crest of the chalk downs along the northern edge of the AONB, passing several ancient sites, is often claimed to be one of Britain's oldest roads, having been used since prehistoric times. The northern section of the Ridgeway, after having left the North Wessex Downs at Streatley, crosses the River Thames and continues through the Chiltern AONB.

The tracks that now form the Ridgeway National Trail were at one time part of a prehistoric network of routes that stretched across southern England from the Dorset coast to the Wash on the Norfolk coast. Prehistoric man travelled through this area on his way to meet at special gathering sites where goods and animals could be traded, and these tracks also connected many of the iconic sites such as burial mounds and Iron Age hill forts that they left behind. During the Dark Ages invading Saxons and Vikings used the tracks to aid their advance into Wessex, while in medieval times invaders were replaced by drovers driving livestock from the West Country to the Home Counties and pilgrims travelling to important religious sites.

These days it is walkers, cyclists and horse riders who travel along this ancient route and it is hard to beat a visit to the Ridgeway. Sit amidst this ancient landscape, high above, the skylarks sing, far off, a church bell rings, the wind rustles through the trees, sweeping views stretch out over the rounded chalk hills with wide open skies. Let your mind wander and imagine the travellers that have been this way before over the millennia.

Segsbury Camp/Letcombe Castle

When travelling westwards along the Ridgeway from the River Thames at Streatley the first of the sites that the trail passes is Segsbury Camp, or Letcombe Castle (SU384844). This Iron Age hill fort on the crest of the Berkshire Downs, beside the Ridgeway above Wantage, was probably occupied between 600 and 200 BC. The

HISTORY – FROM THE STONE AGE TO THE PRESENT DAY

▲ The earthworks of the former Iron Age hill fort of Segsbury Camp

fort consists of an extensive single ditch and rampart with an entrance on the eastern side; a modern track cuts north–south through the earthwork. Excavations have found fragments of Iron Age pottery and metalwork, and a recent geophysical survey has revealed evidence of several round houses within the structure – hut circles are visible as cropmarks in aerial photographs. The discovery of a burial cist on the south side of the rampart suggests that the hill fort was reused in the Saxon period.

Uffington Castle and White Horse

Five and a half miles to the west along the Ridgeway is Whitehorse Hill, the highest point in Oxfordshire at 856 feet, crowned by the remains of Uffington Castle (SU299863). The 'castle' dates from the Early Iron Age (although excavations have revealed some evidence of earlier Bronze Age activity) and consists of two earth banks separated by a ditch, with an entrance on the eastern side; the inner rampart was formerly lined with sarsen stones.

▲ The famous Uffington White Horse looks out of the Vale of White Horse

▲ Magical Wayland's Smithy – one of several prehistoric sites along the Ridgeway

Just to the north of the hill fort is the stylized outline of the Uffington White Horse. This enigmatic hill figure, carved on the steep scarp face of the downs looking out over the Vale of White Horse, is unique amongst the prehistoric sites that our ancestors have left us. Without doubt it is one of the most mystical features of the North Wessex Downs. There is more information about the Uffington White Horse and the more recent white horse hill figures within the North Wessex Downs in Monuments, White Horses and Crop Circles section on pages 140–43.

Wayland's Smithy

About 1½ miles west of Whitehorse Hill is Wayland's Smithy – an atmospheric Neolithic chambered long barrow faced with large sarsen boulders and surrounded by beech trees. Excavations undertaken in 1963 have shown that the long barrow you can see today actually covers an earlier burial structure that was constructed around 3600 to 3500 BC. The earlier burial, which contained the remains of fourteen people, consisted of a sarsen stone floor and wooden burial chamber covered in earth. Recent dating methods have shown that it is likely that the barrow was used for no more than fifteen years.

The second chambered barrow, constructed some years later, probably around 3450 to 3400 BC, contained the remains of eight people in two chambers.

The barrow, mentioned in a Saxon land charter relating to Compton Beauchamp in AD 955, is named after Wayland, a magical smith in Norse mythology. Wayland was said to own a white horse and the close proximity of the Uffington White Horse may explain the name of this barrow. A local legend states that any traveller whose horse required a shoe should leave it with a coin next to the tomb; on returning the horse would be shod and the coin gone. The legend was referred to in Sir Walter Scott's novel *Kenilworth*:

> …you must tie your horse to that upright stone that has a ring in it, and then you must whistle three times and lay me down your silver groat on that other flat stone, walk out of the circle, sit down on the west side of that little thicket of bushes, and take heed you look

▲ Looking out over the Vale of White Horse from the earthworks of the former Iron Age hill fort of Liddington Castle (see following page)

neither to right nor to left for ten minutes, or so long as you shall hear the hammer clink, and whenever it ceases, say your prayers for the space you could tell a hundred – or count over a hundred, which will do as well – and then come into the circle; you will find your money gone and your horse shod.

Alfred's Castle

To the south of Wayland's Smithy, close to Ashdown House, is the feature known as Alfred's Castle (SU277822), a fairly small earthwork that was probably used as a fortified Iron Age homestead as distinct from the much larger and more strategically sited Iron Age hill forts nearby. The site, known as Ashbury Camp before the eighteenth century, was first occupied from the Late Bronze Age and was also used for a Romano-British farmhouse in the late first century AD.

Some have suggested that Alfred's Castle may have been used as a rallying point before the Battle of Ashdown where Alfred the Great defeated the Danes in AD 871; Saxon artefacts found in the area give some credence to the idea. However, as to where the battle actually took place, opinions are split; some suggest the downs above Ashbury while others suggest an area further east around Aston Upthorpe Downs or Blewburton Hill.

Lambourn Seven Barrows Cemetery

Throughout the North Wessex Downs the ground is littered with the remains of ancient burial mounds, none more so than around Lambourn Seven Barrows Cemetery (SU327826). Situated just to the north of Lambourn, alongside the Lambourn to Kingston Lisle road between Postdown Farm and Sevenbarrows Farm, this Bronze Age barrow cemetery contains a large number of barrows. Many of these exist as earthworks, while others, having been destroyed by ploughing, can only be identified from cropmarks.

The core of the cemetery comprises ten barrows arranged in two parallel north-west to south-east rows, with the remaining barrows dispersed over a wider area. Various types of barrow are represented, including bowl barrows which look like an upturned bowl; bell barrows that have more of a bell-like shape, with a flat platform and encircling ditch; and disc barrows that have a flat circular area, surrounded by both a ditch and a bank, with a small hump of earth in the centre marking the position of the grave itself.

A much older Neolithic long barrow lies to the north-west (SU322833), though this has been badly affected by ploughing in the past.

Liddington Castle

Continuing west along the Ridgeway brings us to Liddington Castle, with a commanding position high up on the downs looking out over Swindon (SU208979). Construction of the single ditch and rampart started around 800 or 700 BC, although recent excavations have shown that the ramparts were constructed in four phases. Finds at the site have included fragments of pottery from the Bronze Age, Iron Age and Romano-British period. It has been suggested by some that Liddington Castle was the site of the Battle of Mons Badonicus or Mount Badon (circa AD 500) where the Britons (supposedly led by the legendary King Arthur) defeated the invading Saxons. On Liddington Hill, next to the trig point, a view indicator was installed to mark the new millennium so that you can identify various places – the view was much admired by local authors Richard Jefferies and Alfred Williams (see pages 149–50).

Barbury Castle

The final site along the Ridgeway, before it heads south to Overton Hill just east of Avebury, is the imposing Iron Age remains of Barbury Castle (SU149762). Standing on a spur of the downs at a height of 869 feet, the double ditch and earth ramparts offer commanding views. In 1996 a geophysical survey revealed traces of forty hut circles inside the castle. Military use during the Second World War led to the western entrance being substantially widened and to the placing of anti-aircraft guns. The castle is believed to be the site of the Battle of Beranburgh (Beran Byrig) in AD 556 when the Saxons, led by Cynric and his son Ceawlin (who later became King of Wessex in AD 560), defeated the Britons.

Overlooked by the nearby Liddington and Barbury hill forts is the village of Chiseldon and it was here, in 2004, that a unique discovery was made known as the Chiseldon Cauldrons. A 6½-foot-diameter pit containing twelve bronze and iron cauldrons was found; the largest group of Iron Age cauldrons ever to be discovered in Europe. The British Museum, which has one of the cauldrons on display, is continuing to investigate the finds – maybe one day we'll know more about the Iron Age feast and why the cauldrons were buried.

Bincknoll Castle

If we leave behind the Ridgeway and continue in an anticlockwise direction following the edge of the AONB, we reach Bincknoll (pronounced 'Bynol') Castle, just to the north of Broad Hinton in the north-west of the region, overlooking Wootton Bassett (SU107793). This was originally the site of an Iron Age hill fort lying at the end of a triangular-shaped promontory on the escarpment; the fort was later used as the location for a Norman motte and bailey castle (see page 49).

▲ The earth bank and ditch of Barbury Castle

Oldbury Castle

Next stop heading south is Oldbury Castle, sitting on the top of Cherhill Hill (SU049962) in the far north-west corner of the AONB. Dating from the Iron Age and consisting of two concentric banks and ditches (bivallate), part of the southern edge of the earthwork and surrounding area has been damaged by digging for flint. Limited excavations have recovered pottery, quern stones (used for grinding grain to make flour), loom weights, a weaving comb and animal bones. The Lansdowne Monument (see Monuments, White Horses and Crop Circles on page 139) stands in the western corner of the fort.

Oliver's Castle

This fort, which is roughly triangular in shape, stands right on the western fringes of the AONB looking out over Devizes (SU000646). It has a single ditch with rampart and entrance on the eastern side; excavations in the early 1900s revealed fragments of pottery from the Bronze Age, Iron Age and Romano-British period. The name, Oliver's Castle, may have mistakenly arisen from the fact that a Civil War battle occurred nearby, where Oliver Cromwell's Parliamentarian troops were routed by Royalist cavalry, even though Cromwell himself was never there.

Ancient Features overlooking the Kennet and Avon Canal

The remains of a Neolithic causewayed camp at Rybury (SU083639) look out over the Vale of Pewsey. Fragments of a Neolithic bowl have been found here; however, the original Neolithic camp has been obscured by the construction of a later Iron Age hill fort.

Continuing east over Tan Hill and Milk Hill brings us to Knap Hill Camp (SU121636). This is also a Neolithic causewayed camp, similar to that at Windmill Hill near Avebury, consisting of a single ditch following the contours of the hill and dating from around 3500 BC. A later earthwork enclosure lying adjacent to the Neolithic camp was found to contain artefacts from the late Iron Age or early Romano-British period. A short distance to the west, on Walkers Hill, is Adam's Grave (SU112633), the remains of a Neolithic long barrow.

To the south, across the Vale of Pewsey, close to the southern edge of the AONB, is Marden Henge, sometimes referred to as the Hatfield Earthworks (SU090582). Marden Henge is one of Britain's largest, and probably least understood, henge monuments enclosing an area of around 35 acres – dating from the late Neolithic period (2400–2000 BC). Although the earthworks have been damaged by ploughing and general erosion, the irregular earth bank and internal ditch are still visible.

Within the henge, close to the eastern entrance, there used to be a large mound, known as the Hatfield Barrow. This was excavated in the early 1800s; however, following the excavations the structure collapsed and a few years later the resulting mound was levelled, leaving no visible trace. Further excavation in the late 1960s, centred on the northern entrance of the enclosure, uncovered the crouched burial of a young adult female as well as pieces of late Neolithic pottery, flint and animal bone. More recently, in the summer of 2010, English Heritage undertook further investigations at the site and this revealed the remains of a Neolithic ceremonial building along with other finds; maybe one day, once the findings from these investigations have been revealed, we will know a little bit more about Marden Henge.

Head east along the Kennet and Avon Canal and just to the north of Great Bedwyn is Chisbury Camp (SU278659), nowadays bisected by a modern road. This is a fort consisting of two, and in some places three, circuits of defences surrounding the oval-shaped enclosure. Finds from the site have included Bronze Age artefacts and fragments of Iron Age pottery. (Built on the edge of the earthworks is the former St Martin's Chapel – see page 109.)

Ancient Sites of the North Hampshire Downs

The final set of sites lie within the North Hampshire Downs in the south-eastern part of the AONB. Fosbury Camp (SU319565) is a fine example of an oval bivallate (two concentric banks and ditches) Iron Age hill fort. There's not much to be seen on the tree-covered

▲ Fosbury Camp earthworks near Vernham Dean

northern side, but the earthworks on the south-west side, overlooking the valley, are well pronounced, with a commanding view.

At 974 feet Walbury Hill (SU374618) is not only the highest chalk hill in England but it is also crowned by the largest Iron Age hill fort in Berkshire. The earthwork, high up on the scarp of the North Hampshire Downs, consists of a single rampart and ditch with occasional traces of a counter-scarp and two entrances.

Continuing east along the scarp is Beacon Hill (SU457572). This is a well-preserved example of a contour fort with rampart, ditch and counterscarp situated to the south-east of Highclere, with a single entrance on the south-east side, flanked by parallel banks on each side. Traces of hut circles have been identified within the structure and fragments of Bronze and Iron Age pottery have been found; the tomb of the 5th Earl of Carnarvon, who helped discover the tomb of Tutankhamun, stands in the south-west corner of the fort (for more information on the Earl and Highclere Castle, see Stately Homes on page 120). To the south of Beacon Hill is a group of ten Bronze Age barrows, or burial mounds, known as Seven Barrows (SU462553) (not to be confused with the site of the same name near Lambourn); close by is a memorial to Sir Geoffrey de Havilland (see Monuments, White Horses and Crop Circles – page 138).

The last stop on our journey through the region's prehistoric treasures lies to the east of Beacon Hill where, at Ladle Hill (SU748568), are the remains of an unfinished Early Iron Age hill fort. This has allowed archaeologists to understand more about the methods used by our Iron Age ancestors when constructing a hill fort, as it displays features that would not be visible in a completed earthwork, such as the main ditch and rampart, which were started in several locations leaving causewayed sections that would have been removed in the completed structure.

Just to the north of the hill fort is a well-preserved disc barrow measuring 170 feet in diameter.

THE TIME OF THE ROMAN EMPIRE

Following the Roman invasion of Britain in AD 43 the Iron Age was swept aside to be replaced by a Romano-British culture that lasted for almost 400 years. The Romans started to build fortified settlements and a complex network of roads soon started to emerge.

Their largest town in the area was *Calleva Atrebatum* (Silchester) to the south of Reading and just outside the AONB; from here roads headed east to *Londinium* (London) and west to *Aquae Sulis* (Bath) via *Cunetio* (near Mildenhall). Another ran north-west through *Corinium Dobunnorum* (Cirencester) to *Glevum* (now Gloucester) passing Wickham, where Roman coins and pottery have been found – this was known as the Ermin Way. A third route, the Portway, headed south-west to *Sorviodunum* (Old Sarum) passing close to Hannington and St Mary Bourne, while a fourth road headed north through Streatley and Brightwell-cum-Sotwell to Dorchester-on-Thames.

From *Cunetio* a road headed north passing close to Ogbourne St George on its way to *Durocornovium* (near Swindon), while another road headed south-east through Savernake Forest passing Tidcombe: here the current roadway – the Chute Causeway – follows this former Roman road before it continues to *Venta Bulgarum* (Winchester).

Incidentally, the Chute Causeway is claimed to be haunted by the ghostly figure of a former vicar of Chute, who, during the Black Death, is said to have persuaded his sick parishioners to go to a camp on the causeway, where they would be cared for. He never came back and left them to die; unfortunately for the vicar his plan failed as he too died from the plague and now his ghost wanders the road in penance for eternity.

Cunetio, just to the east of Marlborough and located just across the River Kennet from the present village of Mildenhall (sometimes both called and written 'Minal'), is of particular interest. This former Roman fortified town was, as just mentioned, built at the crossing of several roads, but unlike many Roman towns that continued to grow and flourish after the Romans left, *Cunetio* was completely abandoned, with no visible signs that it ever existed except for cropmarks in aerial photographs. Some excavations were undertaken in the 1950s and a small hoard of coins was found in the 1960s. However, it was in 1978 that a vast number of Roman coins, just short of 55,000, were discovered here, making it the largest hoard of Roman coins ever found in Britain. The coins, dating from the third century, are housed in the British Museum, while the pot in which they were found is on display at the Wiltshire Heritage Museum in Devizes. The site was later investigated by Channel 4's *Time Team* in 2009.

The Antonine Itinerary, a register of Roman stations along various roads that was probably first recorded in the early third century, mentions a place known as *Spinis* lying between *Calleva Atrebatum* and *Cunetio*. This may have been located around Woodspeen or Speen just to the west of Newbury; however, no remains have so far been discovered.

Throughout the region, several villas have been identified. At Alfred's Castle near Ashdown (SU277822), the Romans built a villa within the former Iron Age earthworks. Another villa has been located at Starveall Farm near Aldworth (SU259815) and excavations have revealed it to be a fairly high-status building, with hypocaust (underfloor) heating, pavilioned portico and bath complex. Other villa sites have been identified within the Lambourn Downs at both Stancombe Downs and Maddle Farm (SU305822) by the concentration of building material and associated Roman finds located in the ploughed soil. Adjacent to the fairly large agricultural estate at Maddle Farm was the site of a Romano-British settlement at Knighton Bushes.

During the construction of Brunel's Great Western Railway in 1839, a Roman villa was discovered at Lower Basildon, complete with a beautiful mosaic floor. Unfortunately this was completely destroyed by the workmen as they went about building the railway, but not before the antiquarian and archaeologist Charles Roach-Smith (1806–1890) had made some detailed drawings that now form the only record of the mosaic floor.

Following excavations in the 1990s it was confirmed that Lowbury Hill (SU540822) near Aldworth was the site of a former Roman temple that had also been used as a later Anglo-Saxon burial site. A possible second site, identified through aerial images, is located on Churn Hill (SU522844).

The 'Roman star' within the North Wessex Downs can be found in the grounds of Littlecote House close to the River Kennet between Hungerford and Ramsbury (SU300705). Here we have the remains of a fairly large

▲ Part of the Orpheus Mosaic at the former Roman villa beside Littlecote House

▲ Looking east along the Wansdyke – a Saxon earthwork stretching over the Pewsey Downs seen here near Tan Hill

villa complex including gatehouse, workshops, barns and bath house that were occupied from the late first to the fourth century.

The first reference to the Roman site at Littlecote was in 1727 when William George, estate steward to Sir Francis Popham, owner of Littlecote House, first uncovered the Orpheus Mosaic. The mosaic floor, dating from around AD 360, was described as 'the finest pavement that the sun ever shone upon in England'. Despite this, the mosaic was reburied in 1730 and later declared lost. Yet the story does not end there, for in 1977 the mosaic was rediscovered. A full restoration was undertaken to repair the damage that the mosaic had suffered from frost and weathering after it had first been uncovered, with damaged panels being replaced with modern equivalents using the detailed drawings that had been undertaken when originally discovered.

In the centre of the mosaic is Orpheus – musician and priest to Apollo, the sun god – surrounded by four goddesses representing the four seasons. Although called the Orpheus Mosaic, some have suggested that it may be a representation of the pagan cult of Bacchus.

The buildings were abandoned shortly before the Romans withdrew from Britain, and fell into ruin.

The demise of the Roman Empire in Britain, around AD 410, paved the way for the 'Dark Ages' and the rise of the Anglo-Saxons, which lasted until the Norman invasion of England in 1066.

SAXON EARTHWORKS

Little remains in the region from the early Saxon days save for a massive linear east–west earthwork known as the Wansdyke, traversing the Marlborough Downs above the Vale of Pewsey from Marlborough to Morgan's Hill. Fortunately little damage has been done to it over the centuries and it is clearly traceable on the ground as an earth bank up to 13 feet high in places with a 6½-foot-deep ditch. The earthwork is especially well defined around Milk Hill and Tan Hill above All Cannings (between SU080650 and SU110645), and offers sweeping views along its length looking to the east and west.

Its origins, like many ancient structures, are unclear, yet archaeological data has shown that it was probably built during the fifth or sixth century. In other words it was built sometime after the Roman withdrawal, but before the complete takeover by the invading Anglo-Saxons. Since the ditch is located on the north side, it may have been constructed as a defence against West Saxons encroaching from the upper Thames Valley, or maybe its purpose was one of demarcating territorial areas; opinion is divided.

Another reminder from Saxon times is the *herepath*, or Green Street, that crosses the Ridgeway near Avebury. A *herepath* is an army road or path (*here* meaning 'armed host') that was used to move armies around the country more easily.

MEDIEVAL CASTLES AND CIVIL WAR

Within the North Wessex Downs and its immediate area the Normans built a number of motte and bailey castles, and the earthworks of these early fortified structures can still be seen today.

Let us start our exploration of the medieval castles up in the far north-east corner of the North Wessex Downs at South Moreton (SU557880); incidentally the name is said to be derived from the Anglo-Saxon word for 'town on the moor'. Here, just west of the Church of St John the Baptist, is a large, roughly circular mound or motte standing 13 feet high and 160 feet in diameter, surrounded at its base by a broad, deep ditch.

Lying just to the north-east of the AONB is Wallingford (SU609896). It wasn't until Alfred the Great made it one of his fortified towns or burhs, that Wallingford became important; the extensive earth banks and ditches are still clearly visible. Following the Norman Conquest, a castle was built, which came to dominate the town's history for the next 600 years before being destroyed by Oliver Cromwell at the end of the Civil War; Wallingford had been a Royalist stronghold.

To the south-west from here is Hampstead Norreys, tucked amongst the downs just to the north of the M4. In the woods on the south side of the village are the remains of a small motte (SU528760) although, because of its small size, there is some debate as to whether this mound is actually the remains of a Norman castle as the keep would have been rather small.

Right in the far north-west of the AONB, high up on the crest of the chalk ridge, is Bincknoll Castle (SU107793), just to the north of Broad Hinton. Here, the Normans re-used the former Iron Age fort as the basis for a motte and bailey castle. The castle may have been constructed by Gilbert of Breteuil, who acquired a number of manors around Broad Hinton after the Norman Conquest. Unfortunately, the motte, which measures over 165 feet in diameter by 10 feet high with a 6½-foot-deep ditch, has been badly damaged by later quarrying.

We have already learnt earlier in this chapter that the Marlborough Mound (SU183686) – no public access – was used as the basis of a Norman castle, although the mound itself is a contemporary of Silbury Hill. To the east along the Kennet Valley is Hamstead Marshall. Here, close to St Mary's Church and the site of Hamstead Marshall House (see page 119), are the remains of three mottes overlooking the River Kennet. The motte nearest the road (SU421669) is about 200 feet in diameter with a height of 22 feet. Just 400 feet away is a slightly smaller motte (about 165 feet in diameter and 15 feet high) which is also surrounded by a ditch with an adjoining square bailey; the third lies half a mile to the east (SU429665).

These castles were probably first constructed in the late eleventh or early twelfth century by either Hugolin the Steersman or John Marshal; certainly there were buildings here in the early thirteenth century when Henry III stayed. Some have suggested that the castle mentioned as being 'at Newbury' in the mid-twelfth-century wars between Empress Matilda and King Stephen was actually the castle at Hamstead Marshall.

Following the death of Henry I in 1135, the throne of England was claimed by a cousin, Stephen of Blois. However, Henry's daughter, Empress Matilda, did not let the situation go unnoticed. In 1139 she landed in England and fifteen years of civil war ensued – a time that became known as the 'Anarchy' where many of the local castles (including South Moreton and Ludgershall) were besieged at various times. Matilda's son, Henry of Anjou, took up the cause and following the siege of Wallingford, Stephen and Henry agreed a peace on the

▲ The remains of Donnington Castle, having been destroyed after the Civil War

understanding that Stephen recognized Henry as his heir; on Stephen's death in 1154 Henry II came to power.

To the north-east of Hamstead Marshall is Donnington Castle (SU460691), which lies just outside the AONB to the north-west of Newbury. It was held by the Abberbury family from the late thirteenth century and, in 1386, Richard II granted Sir Richard Abberbury a licence to fortify the castle (Sir Richard had fought alongside Richard II's father, Edward the Black Prince, at the battles of Crécy and Poitiers during the Hundred Years' War between England and France).

The castle consisted of a curtain wall with four round corner towers, two square-walled towers and a substantial gatehouse. In the early fifteenth century the castle was held by Thomas Chaucer, son of Geoffrey Chaucer, the great poet and author of *The Canterbury Tales*, and it later passed into the ownership of the Crown. During the Civil War the castle remained a Royalist stronghold enduring a twenty-month siege before finally being surrendered to the Parliamentarians when the Royalist cause appeared lost. Parliament then voted to demolish the castle; only the striking twin-towered fourteenth-century gatehouse is now left standing.

To the south of the AONB, not far from Andover, is Ludgershall Castle (SU263511), built as a medieval fortress surrounded by earth banks and ditches in the late eleventh century. The castle was most likely built by Edward of Salisbury as he is mentioned in the Domesday Book (1086) as holding Ludgershall at that time. However, on his death it passed to Henry I. Around the time of Henry III, it was used more as a hunting lodge surrounded by parkland, he also made several changes to the castle including the building of a new great hall in the mid-thirteenth century. By the early fourteenth century, Ludgershall was known as 'the king's manor' yet, by the early sixteenth century it was in a ruinous state; all that remains now are some ruined walls and a corner tower.

Another former motte and bailey castle, which again lies just outside the AONB, was built at Devizes (SU002613). The original wooden castle was built in the late eleventh century by Osmund, Bishop of Salisbury, although this only lasted around thirty years before being rebuilt in stone. During the Civil War the castle, which was a Royalist stronghold, suffered numerous attacks by Parliamentary forces, until 1645 when it was surrendered to Oliver Cromwell; like Donnington Castle, the castle at Devizes was destroyed by order of Parliament. During the nineteenth century, the Leach family built the present Victorian 'castle' (which is private).

The English Civil War of the mid-seventeenth century consisted of a series of armed conflicts between Parliamentarians and Royalists. The Battle of Roundway Down (SU020650) was fought on 13 July 1643 just to the north of Devizes; the only registered battlefield site within the North Wessex Downs. The Parliamentarians under Sir William Waller, who were besieging Devizes after a battle at Bath, had to withdraw to Roundway Hill to face a smaller Royalist force under Lord Wilmot. Despite being outnumbered, the Royalist forces won the day, giving them a temporary dominance in the south-west.

Newbury, probably as a consequence of its location on the route from London to the west, witnessed two battles during the Civil War. The First Battle of Newbury occurred on Wash Common in 1643, when Royalists tried to prevent the Earl of Essex and his men returning to London after the Siege of Gloucester. Just over a year later, the Second Battle of Newbury was fought at Speen just to the south of Donnington Castle, securing a victory for the Parliamentarians.

The late medieval period brought with it more than just castles; many of the churches within the North Wessex Downs have their roots in the Norman period. Some notable Norman churches include the Church of St Thomas at East Shefford and the Church of St Mark and St Luke at Avington, whereas at St Swithun's Church in Wickham the rare late Anglo-Saxon tower remains. We will learn more about these churches, as well as stately homes, in Chapter 5. More modern developments such as the Kennet and Avon Canal and the railways are covered in Chapter 3, while some of the towns and villages within the North Wessex Downs are explored in Chapter 4.

North Wessex at War

The North Wessex Downs has seen its fair share of service during both the First World War (1914–18) and the Second World War (1939–45). The First World War saw the establishment of the Chiseldon Camp (see below for more details), and later numerous airfields in the region played an important role in the transportation of troops during the 1944 D-Day campaign of the Second World War. Although little remains on the ground, many of the former airfields are still clearly visible in aerial photographs (aerial views can be examined on Google Earth). There were also large ammunition depots, and decoy systems were established to try to lure enemy planes away from important targets. The threat of invasion early on in the Second World War led to the Kennet and Avon Canal becoming one of many Stop Lines, reinforced with pillboxes and anti-tank defences, manned by the Home Guard, with the aim of hindering any German invading forces.

Even stately homes, such as Basildon House, played their part in the war effort, being requisitioned as military bases, especially on the run-up to the D-Day campaign. Littlecote House was requisitioned by the US 101st Airborne Division as office space and sleeping quarters for 506th officers; other ranks lived in Nissen huts built alongside the main drive between the house and the east lodge. Colonel Robert Frederick Sink (506th Regimental Commander of the Parachute Infantry Regiment) used the library as his office and because of this connection a memorial plaque was unveiled on the sixtieth anniversary of 'Operation Market Garden'.

To find out more about the wartime history of the Kennet Valley, visit the Kennet Valley at War Trust Museum at Littlecote House. The house, now owned by Warner Holidays, has kindly set aside a room for use by the trust to house a small museum, which is open for the public to visit.

CHISELDON CAMP

It may come as no surprise that Chiseldon Camp (SU186780), established in August 1914 after the outbreak of the First World War, was originally known as Draycot Camp, as it was located right next door to the little hamlet of Draycot Foliat. The site was chosen not only because of the road and rail links (the Midland and South Western Junction Railway – M&SWJR – passed close by), but also because the area around Burderop Park already had military connections, having been used for annual camps and manoeuvres by the Wiltshire Yeomanry. Burderop Park, situated between Wroughton and Chiseldon and owned by the Calley family for over 200 years, was used in both world wars; during the Second World War it was home to a US military hospital.

Chiseldon Camp was primarily used for training infantry, with upwards of 10,000 men in the camp at any one time; the training of infantrymen took just

▲ Sarsen stone and plaque commemorating Chiseldon Camp

fifteen weeks – much shorter than the more typical twelve to eighteen months allowed in peacetime. The surrounding downland was extensively used for training, with grenade and rifle ranges below Liddington Hill, including targets and firing points at various distances, along with an extensive trench system constructed near Lower Upham Farm.

In 1915 a hospital was built to treat wounded soldiers and the camp even gained its own railway station on the line between Swindon and Marlborough (part of the M&SWJR, which had opened in 1881), known as the Chiseldon Camp Halt.

After the Armistice, the camp became a demobilization centre and a holding camp for Commonwealth troops awaiting ships for home. In 1922 parts of the camp were sold at auction, with the remainder being occupied by the School of Military Administration. Some five years later the Army Vocational Training Centre was established at Chiseldon to train long-serving soldiers in readiness for a return to civilian life.

A nearby farm was leased and greenhouses and workshops were built to allow a range of various skills to be taught. During the 1930s, produce from the farm was used in the army kitchens, with any surplus being sold at market.

However, the outbreak of the Second World War in 1939 brought a sudden change to Chiseldon with the arrival of the Motor Training Battalion of The King's Royal Rifle Corps (KRRC), one of the first motorized regiments of the British Army. In 1943 the KRRC left for Yorkshire and the camp came under the control of the US Army for the build-up to the invasion of Europe.

For the next two years both Chiseldon and its near neighbour at Ogbourne St George were used to train many thousands of US soldiers who had arrived in Britain, although the first arrivals at Ogbourne were actually Canadians in late 1940; almost no trace of the camp at Ogbourne St George remains.

Following the end of the war the camp returned to British control, and was again used to prepare soldiers

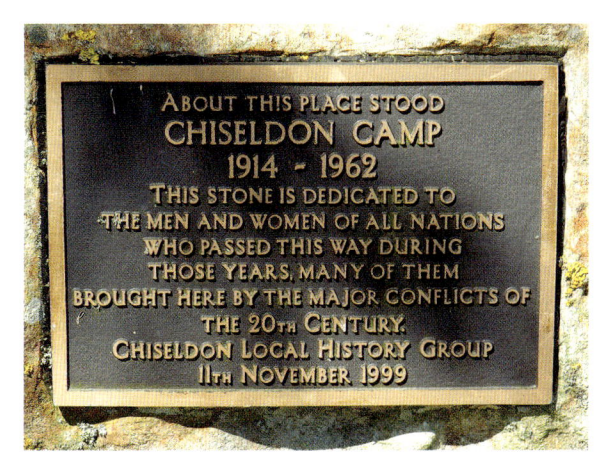

▲ The memorial plaque commemorating the men and women who passed through Chiseldon Camp

▲ Former Second World War pillbox in the Vale of Pewsey, not far from the Kennet and Avon Canal

for their return to civilian life as well as running secondary education training of young Polish troops who had been fighting alongside the Western Allies.

By 1948 Chiseldon had become a transit camp for units between overseas postings and, at the end of National Service, the camp finally closed.

The camp was mostly demolished in the 1970s, with little remaining today, except for the old railway platform and a few roads. In 1999 a memorial was unveiled by the Chiseldon Local History Group to commemorate the camp's existence at the junction of Ladysmith Road and Sambre Road (SU186780).

GENERAL IRONSIDE AND THE KENNET AND AVON CANAL

> 'We shall fight on the beaches, we shall fight on the landing ground, we shall fight in the fields and in the streets, we shall fight in the hills, we shall never surrender.'
> WINSTON CHURCHILL, *June 1940*

Following the defeat at Dunkirk in 1940, Britain was faced with the threat of a German invasion and, to counter this, Winston Churchill put General Sir Edmund Ironside, Commander-in-Chief Home Forces, in charge of organizing Britain's defences.

Hitler gave the go-ahead for preparations to be made for the possible invasion of Britain – Operation Sea Lion – and throughout the summer of 1940, while the Battle of Britain raged in the skies, Britain responded by hastily building a complex series of defences including 18,000 pillboxes. The primary purpose of these defences was to delay German armoured columns from advancing as easily as they had done in France.

In addition to the main coastal defences, a series of barriers, known as General Headquarters Anti-Tank Lines or 'GHQ Stop Lines', were devised. These utilized both existing natural obstacles such as rivers and marshland and also man-made ones including canals and railway embankments, supplemented by the aforementioned pillboxes, gun emplacements, anti-tank ditches, and other obstacles.

To construct and man these defences, Anthony Eden, Secretary of State for War, ordered the formation of the Local Defence Volunteers (LDV), comprising men who were too old to join the regular army or those in protected trades and thus exempt from conscription. Following a speech by Churchill the volunteers became known as the Home Guard – the real 'Dad's Army'.

The Kennet and Avon Canal, which slices across southern England, became GHQ Stop Line Blue stretching from Semington near Bradford-on-Avon to Theale, just west of Reading. Here it joined the GHQ Stop Line Red that headed north along the Sulham Gap (through which the River Pang flows) to Pangbourne and then went upstream along the River Thames towards Abingdon. Pillboxes and anti-tank gun emplacements were strategically sited along the whole length of these stop lines, and concrete obstructions were placed across bridges.

Several pillboxes that formed part of Stop Line Blue can be seen at Dun Mill Lock (SU350681) on the Kennet and Avon Canal just to the east of Hungerford, with a good selection along the Sulham Gap towards Pangbourne, especially around Sulham village (SU643741) and alongside Nunhide Lane.

Fortunately, Hitler's Operation Sea Lion was postponed, the imminent threat of invasion receded and the pillboxes and assorted defences were never used in combat.

RAF STATIONS AND AIRFIELDS

RAF Alton Barnes

In the south-west corner of the AONB in the Vale of Pewsey is Alton Barnes, once home to RAF Alton Barnes (SU104618) which was operational from 1935 until 1945. The airfield was used for pilot training, which included touch-and-go landing practice, as well as glider training. Later Alton Barnes was upgraded to a Relief Landing Ground, complete with hangars, men's quarters, a watch tower and an anti-aircraft gun station on Milk Hill.

As with most of the Second World War airfields, little remains visible today, although there are two memorial plaques. This first memorial, located beside the Kennet and Avon Canal to the south of the village (SU107613), is dedicated to the pilot and navigator of a twin-engined Albemarle Bomber V1755 from RAF Keevil near Trowbridge. On 25 October 1944 the Albemarle was towing a glider when disaster struck, with the glider overtaking the towing plane and pulling its tail to a critical angle, causing the plane to crash. Fortunately the glider managed to recover and landed safely at Alton Barnes. The second memorial, located above the entrance to the only remaining air raid shelter beside the road that runs between Honeystreet and Alton Barnes, commemorates the former RAF site and the RAF personnel who were killed while stationed there (SU104618).

▲ Memorial plaque on the last remaining air raid shelter at former RAF Alton Barnes

RAF Clyffe Pypard

Just to the south of Clyffe Pypard village, located high up on the flat downland above the chalk escarpment, was RAF Clyffe Pypard (SU070760). The airfield opened in 1941 with grass runways and was used by the Elementary Flying Training School (EFTS) for pilot training, flying the de Havilland Tiger Moth (RAF Alton Barnes to the south was used as an alternative landing ground). The site was used until 1961 as a transit camp before reverting back to farmland, though a few buildings remain including a hangar and pillboxes.

RAF Hampstead Norreys

RAF Hampstead Norreys – sometimes known as Hampstead Norris – was located to the east of the village of Hampstead Norreys, adjacent to the present Haw Farm (SU544770). The airfield, a satellite for the main airfield at Harwell, was operational from 1940 until 1945, and was originally used as an RAF Bomber Command Operational Training Unit (OTU) where pilots were trained to fly Wellington Bombers. The

▲ Memorial plaque at the end of the main runway at the former RAF Harwell airfield

airfield was attacked on several occasions, the worst being in 1941 with the Luftwaffe dropping high explosive bombs and multiple incendiaries. During the later stages of the war the airfield was used extensively as a glider training base in preparation for D-Day.

Little remains of the former airfield, except for a farm track at Haw Farm that follows the line of one of the runways, and a few pillboxes and brick air raid shelters in the nearby woods (private). Yet the location still has links with flying; light aircraft still use the grass landing strip and close to the west side of the site is a flat circular aerial structure. This is a Doppler VHF Omni-Range radio beacon that marks the centre line of the flight path for outbound planes from Heathrow Airport.

RAF Harwell

RAF Harwell (SU483863) opened in early 1937 and was the base for Wellington-equipped Bomber Command Operational Training Unit No. 15 between 1940 and 1944. In the late 1930s Harwell became the location of a prototype pneumatic catapult system designed to assist the take-off of bomber aircraft. The structure consisted of a turntable approximately 100 feet in diameter and two launch trenches about 280 feet long; unfortunately the system was never successfully used to help launch an aircraft.

It was from Harwell during the early years of the Second World War that Wellington bombers left for bombing raids over Bremen, Essen and Cologne. However, in preparation for D-Day, the airfield was given over to Albemarles and later to Stirling glider tugs.

On the night of 5 June 1944, aircraft from No. 38 group of the RAF took off with troops of the 6th Airborne Division and they became the first British soldiers to land in Normandy in the main assault for the liberation of Europe. A memorial dedicated to the men is located at the east end of the main runway from which the planes took off – the memorial is just off the A4185 near the bus stop after leaving the A34 (SU483863).

After the war, the site was taken over by the Ministry of Supply and became the Atomic Energy Research Establishment (AERE), the UK's centre for research and development into civil nuclear power. At one time Harwell had five research reactors, including the first nuclear reactor in Western Europe known as GLEEP – the Graphite Low Energy Experimental Pile, which operated for forty-three years. The site, now renamed Harwell Oxford, continues with cutting-edge scientific research projects (see From the Land to Science – page 70).

RAF Membury

RAF Membury was located to the north-east of Hungerford where the Membury Services on the M4 now stand (SU6307754). The airfield opened in August 1942, originally intended for use by RAF Bomber Command; however, it was redesignated USAAF Station 466 and was used by the USAAF for photo-reconnaissance operations before being later moved to Middle Wallop in Hampshire.

By the end of 1943 the site had become the home of the 6th Tactical Air Depot (TAD) which specialized in the checking and updating of newly arrived P-47 Thunderbolts before they were allocated to operational squadrons.

Following its use during the D-Day campaign, transporting troops in Horsa gliders towed by C-47s, the airfield was returned to the RAF in 1945 and closed in 1946. Some of the former hangars and associated buildings remain and are now seeing a new lease of life as industrial units.

Fans of the BBC TV series *Dr Who* might be interested to know that the former airfield was one of the locations used in *Planet of the Spiders* (1971), where the Doctor takes off in an autogyro; the hangar of the Campbell Aircraft Company which developed and built autogyros, including the single-seat Cricket in 1969, was used in *The Daemons* (1971) as Bessie's garage.

RAF Ramsbury

Located to the south of Ramsbury just across the River Kennet, about 4 miles east of Marlborough (SU270703), RAF Ramsbury became operational in 1942 and was allocated to the USAAF as Station AFF-469 for use by transport and observation squadrons, being one of thirteen bases provided for this purpose. At times throughout its four years of service the airfield was also used by the RAF for pilot training. Ramsbury also had its own Women's Auxiliary Air Force (WAAF) contingent who undertook many jobs including repacking parachutes.

Shortly after the end of the war, Ramsbury was returned to RAF control and continued to operate as an active airfield until the end of March 1946, when its status was downgraded to 'care and maintenance' before final closure in 1947. The site was returned to agricultural use and by the mid-1960s most of the concrete runways had been removed. Today only the outlines of the main runways can be seen in aerial photographs.

RAF Welford

RAF Welford, which opened in the summer of 1943 and is still an active site, lies to the west of Newbury just north of the M4 and the village of Welford (SU410740) and was originally planned as an Operational Training Unit (OTU) airfield. However, a few months after opening the airfield was used by both the RAF and USAAF (Station 474) primarily as a transport airfield, later moving large numbers of troops in support of the D-Day Landings at Normandy (June 1944). Aircraft from Welford also took part in 'Operation Dragoon' – the invasion of Southern France in August 1944 – and 'Operation Market Garden' – the invasion of Holland in September 1944. Following the end of the war, the airfield was mothballed from 1946 until 1955 when, as a result of the Cold War, the station was reopened as a munitions depot for the United States Air Force.

In 1991, Welford was again used to stockpile vast quantities of munitions in readiness for the Gulf War. In 1999 the site was handed over to the MoD Defence Munitions Agency and the USAF is now the sole user; today, Welford is one of the largest heavy munitions compounds for the USAF in Western Europe.

Originally munitions were transported to and from Welford by rail; however, the rail link was abandoned in 1973 in favour of the newly opened M4 motorway; anyone driving east along the M4 between junctions 14 and 13 might catch sight of an access road and distinctive red-bordered 'Works Unit Only' signpost that leads to the site.

RAF Wroughton

Located to the south of Swindon, RAF Wroughton (SU138786) was operational from the late 1930s through to the 1970s, during which time it served as host to aircraft maintenance units. Wroughton proved to be a valuable asset during the war years, undertaking repairs, servicing and modifications on many different types of aircraft. By 1941 the site was also being used for packing aircraft into large crates for shipment overseas and, during late 1943, Wroughton became an assembly point for many of the wooden-framed Airspeed Horsa gliders that played a key role in transporting troops for D-Day in 1944.

After the war, Wroughton continued as a maintenance site and later worked on jet aircraft such as the Gloster Meteor and English Electric Canberra bomber. Wroughton was also the site where many old planes met their fate on a scrapheap. In the 1960s, Wroughton started work on servicing helicopters, and in 1972 the site passed to the Royal Navy as they had largely taken over responsibility for servicing all military helicopters; the site closed in 1978. Ownership of the site then passed to the British Science Museum in 1979 and since then the six large hangars have been used as the museum's large object store, library and archive.

The air base was also home to the RAF Princess Alexandra Hospital, originally known as the RAF General Hospital, which opened in 1941; the 'Princess Alexandra' prefix was added in 1967. The hospital closed in 1996 and has since been redeveloped for housing as Alexandra Park.

RAF Yatesbury

The Royal Flying Corps (RFC), the air arm of the British military, started to train pilots at Yatesbury Field, located on the north side of the A4 beside the minor road to Yatesbury (SU060701), in 1916, originally using Avro 504A and Bristol Scout D biplanes. In 1918 the Royal Flying Corps was combined with the Royal Naval Air Service (RNAS), the air arm of the Navy, to form the Royal Air Force (RAF).

Following the end of the First World War, Yatesbury closed in early 1920 and reverted back to farmland. However, because of rising fears about Germany, the site was reopened in 1936 as an Elementary and Reserve Flying School set up by the Bristol Aeroplane Company (BAC) which had been operating a training school at Filton since 1923; training was carried out with Tiger Moth aircraft. At the outbreak of the Second World War in September 1939, pilot training was transferred to other stations and Yatesbury became a training centre for airborne wireless operators; this later included radar training.

Yatesbury continued to operate as a training centre for radar operators, mechanics and fitters and became the headquarters of 27 Group Technical Training Command between 1954 and 1958, when it was known as RAF Cherhill. The site closed in 1965 after which the land was returned to agricultural use. Some of the former buildings remain and at the time of writing plans have been submitted to develop one of the remaining hangars (Hangar 45), to the west of the road to Yatesbury, into living accommodation.

Flying still lives on at Yatesbury Field as this is now home to the Wiltshire Microlight Centre; at the entrance are two memorial plaques commemorating the former RAF site.

Just to the north of Yatesbury was RAF Townsend, which was originally used as an emergency landing practice area for the flying school at Yatesbury. However, when Germany started bombing airfields during the Second World War it was soon realized that aircraft parked around the airfields could easily be damaged, so a series of Satellite Landing Grounds (SLG) were found. Townsend was one such site (No. 45 SLG) that was used from 1940 to 1944 for storing aircraft. In 2007 a plaque commemorating the site was placed at the entrance to the field (SU064725).

Just to the west of Cherhill (and a stone's throw outside the AONB) was the site of RAF Compton Bassett (now known as Lower Compton). Like its neighbour at Yatesbury, Compton Bassett was an important Radio and Radar Training School from 1940 until it finally closed in 1964 (SU021705).

▲ Memorial on Walbury Hill to the men of the Parachute Regiment who trained there during the Second World War

OTHER LOCATIONS OF MILITARY SIGNIFICANCE

On Liddington Hill, a short way north of the Ridgeway (SU215801), is an old concrete bunker – all that remains of a control bunker for a Second World War 'Starfish' bombing decoy site (the name 'Starfish' came from the code name for one such site which was derived from the original code, SF, for 'Special Fire'). This was one of two sites (the other was at nearby Badbury) commissioned in early 1941 to control fires which would have acted as a decoy to enemy planes targeting Swindon – especially the railway works – just to the north. In the event, large-scale attacks on the area never came and the decoy was never required.

Throughout the Second World War, Savernake Forest was used as an ammunition dump by both the British and US Armies. Between 1940 and 1942 the site was occupied by the British Army and at times over 20,000 tons of ammunition was stored there. From late 1942 until 1945 the forest came under the command of the US Army's Services of Supply and, like the British Army, they continued to use the depot as an ammunition dump. After the war the site reverted back to the British Army and it was used to store ammunition brought back from the Continent before most of it was dumped at sea.

On the northern slopes of Walbury Hill, near the car park (SU370620), stands a small memorial and plaque. This commemorates the fact that the area below was used in 1944 by the 9th Battalion, the Parachute Regiment, in preparation for the successful assault on the German coastal artillery battery at Merville, France, before the invasion of Normandy.

Heading east along the Ridgeway, just after the A34 underpass (SU492834), is a small memorial on the left to Hugh Frederick Grosvenor, a 2nd Lieutenant in The Lifeguards, who lost his life there in an armoured car accident on 9 April 1947, aged just 19.

During 1944 Hampstead Park, which lies just outside the AONB, was home to many US soldiers from the 501st Parachute Infantry Regiment in preparation for D-Day. Shortly before the invasion, General Eisenhower visited the troops encamped in the park. A memorial was unveiled in June 2004 to commemorate the American troops who had camped here 60 years earlier (SU340657).

CHAPTER THREE

North Wessex at work

The landscape of the North Wessex Downs has seen a fair number of changes over the centuries brought about by man's use of the environment in which he lives. The rolling chalk downs have been used to rear vast numbers of sheep and the wool they produced was used in a rapidly growing textile industry during the Middle Ages. Later, the invention of modern farm machinery brought about the intensification of arable farming and the removal of hedgerows and field boundaries, creating the large open landscape we can see today.

The rivers were used to power mills for grinding corn, fulling woollen cloth, paper-making and even making iron for use in agricultural tools. Later, the clear waters of the chalk streams were used for growing watercress, an industry that still survives today on the Bourne Rivulet.

The nineteenth century brought about increased communication through an otherwise isolated area, with the construction of turnpike roads and the Kennet and Avon Canal. This was rapidly followed by the arrival of the railway. Communications were further improved in the late twentieth century with the M4 motorway cutting through the region from east to west and the A34 from north to south. Economic growth and the end of petrol rationing in 1950 led to a rapid increase in car ownership and an emerging motorway network coupled with improved A-roads allowed the road haulage industry to continue growing to meet demand. Plans to modernize the rail system failed to stem the rising losses and the infamous Beeching Report brought about the closure of many rural railway lines and stations during the 1960s, including some of the lines that passed through the North Wessex Downs.

While road traffic levels have continued to climb, there has also been a growth in the number of people using the rail network since the mid 1990s and it is predicted that more long-distance freight will switch over to the rail network.

Tourism is a growing industry in the North Wessex Downs; visitors come to the area to see the many archaeological sites (which were covered in Chapter 2), to visit famous stately homes (see Chapter 5), to enjoy the peace and tranquillity of the open countryside and to wander, on foot, bicycle or horse, along many miles of rights of way.

One industry that doesn't often get a mention is smuggling. More often than not, this is thought of as a coastal activity, yet the very isolation of the North Wessex Downs, coupled with its network of little-used trackways, made it ideal for transporting contraband from the south coast to the more populated regions of Britain. In 1874 the Revd A. C. Smith talked of entire villages in parts of the North Wessex Downs where the main source of employment was smuggling, and it was this smuggling activity that gave rise to the legend of the Wiltshire 'moonrakers'.

There is some debate about the 'moonraker' legend and where it originated; probably the more reliable version is relayed on a memorial stone in Devizes. Smugglers from Bishops Cannings (whether that is the village itself or the parish, which at one time included Devizes, is unclear) had concealed kegs of brandy in Cramer Pond in Devizes. At night, by the light of a full moon, the men were in the act of retrieving the barrels when they were observed by excise men. When challenged about their activity, the quick-thinking locals said they were trying to rake in the big round of cheese, pointing to the reflection of the moon on the shimmering surface of the pond. Fortunately for the smugglers, the excise men rode off laughing at the idiocy of the locals, who were left to retrieve their prized goods. The moral being that Wiltshire folk are both quick-witted and resourceful, and many locals are proud to call themselves 'moonrakers'.

From the Land to Science

SHEEP AND CROPS

For thousands of years, ever since our prehistoric ancestors started clearing the forest following the end of the last Ice Age, man has been using the downs for grazing and crop-growing. Today the North Wessex Downs is essentially an agricultural landscape.

Modern farming has produced an open arable landscape, often with little in the way of either hedgerows or tree cover – although in more recent years an increasing awareness of the importance of wildlife has seen a shift towards replanting hedgerows and setting aside areas for nature conservation.

Approximately 84 per cent of the AONB is classed as farmland with 60 per cent under arable production. The most common crop is wheat (52 per cent), followed by winter and spring barley (22 per cent), and oilseeds such as rape and linseed (11 per cent), the remainder being classed as set-aside. Grazing land accounts for a quarter of the farmed area, with dairy cattle grazing concentrated in the vales and river valleys, while beef cattle and sheep graze the rolling chalk downs. However, there has been a steady decline in the number of livestock within the AONB.

The North Wessex Downs has around 1,000 farms, with just over a third of these being under 50 acres in size; these farms account for only 2 per cent of the land area within the AONB. However, another third of the farms are over 250 acres in size and these account for close to 90 per cent of the farmed area; the average size of a farm within the North Wessex Downs is considerably larger than the national average.

At the time of writing, in the region of 5 to 6 per cent of the economically active population within the AONB are employed in farming and despite the total number of full-time farmers being fairly constant there has been a decline in the number of full-time agricultural employees.

Despite the low number of people employed, farming is an important source of income for the rural economy, with over half of the farm-gate income coming from arable crops, and the remainder, in decreasing importance, coming from dairy, pigs, beef and sheep, and poultry.

East Ilsley, sometimes known as Market Ilsley, was once famed for its sheep fairs. The village had been an important trading centre since the thirteenth century and East Ilsley's Fair was established through a charter granted by Henry III, a second Royal Charter being granted by James I in 1620. The sheep fairs were held from April to October, with tens of thousands of sheep being sold each day. Yet these were not just sheep fairs; cattle and wheat would also have been traded, along with domestic items such as pottery and clothes.

In the late nineteenth century, Compton railway station on the former Didcot, Newbury and Southampton Railway became an important centre for the transportation of sheep to and from the fair at East Ilsley.

While on the subject of agriculture we should not forget a certain Jethro Tull – no, not the British rock group, but the famous inventor and pioneer of agricultural reform known as 'the father of the English Agrarian Revolution'.

Born at Basildon in Berkshire, close to the River Thames, in 1674, Jethro, who originally studied law, decided to try his hand at farming and spent time at both Howbery Farm (16–19 High Street) in Crowmarsh Gifford, where there is an official blue plaque, and at Prosperous Farm just south of Hungerford where he died in 1741; he is buried at Lower Basildon.

At the time, seeds were distributed into furrows by hand, giving heavy sowing densities that were not very efficient, so he invented a machine to do the work instead. He designed his horse-drawn seed drill with a rotating cylinder that had grooves cut into it allowing the seed to pass from the hopper above to a funnel below. They were then directed into a channel dug by a plough at the front of the machine, then covered by a harrow attached to the rear. This reduced the amount of wastage by sowing the seed with precision rather than being cast haphazardly by hand.

He also invented the horse-drawn hoe and changed the design of a plough, with blades set in such a way that grass and roots were pulled up and left on the surface to dry and in 1731 he published his book *The Horse-Hoeing Husbandry*, detailing his system and machinery.

The continuing mechanization of farming, especially the introduction of the horse-powered threshing machine, reduced the need for agricultural workers and led in part to the 'Swing Riots' in the 1830s. Within the North Wessex Downs these became known as the 'Kintbury Riots' as the perpetrators were arrested in Kintbury and all but one of them were transported to Australia; unfortunately, William Winterbourne was hanged for his part in the riots.

FANCY A DRINK?

Throughout the North Wessex Downs there are a number of small, but well regarded, producers of real ale, cider and, to a lesser extent, wine.

West Ilsley was the original home of the Morland Brewery, founded when farmer John Morland set up his brewery in 1711. However, in 1887, the brewery relocated to Abingdon and it was bought by Greene King in 2000. The Morland tradition still survives, however, with brews such as Morland's Original and Old Speckled Hen, the latter being first brewed to commemorate the fiftieth anniversary of the MG car factory in Abingdon.

Butts Brewery, based in Great Shefford, started brewing in 1994 and became the first commercial micro-brewery in West Berkshire since the Phoenix Brewery of Newbury closed down in 1923. In 1995 the West Berkshire Brewery started out in an old converted stable block behind the Pot Kiln pub at Frilsham. The original brewery was a small five-barrel plant – a brewing barrel being thirty-six gallons, which equals four firkins that go to the pub. Within a couple of years the brewery moved to the Old Bakery in Yattendon where they are still happily brewing away.

Honeystreet Ales (brewed at the Stonehenge Brewery in Netheravon, just to the south of the AONB) brew several ales, as well as using apples and pears grown on the Hampshire Downs around Selborne to produce cider and perry. Being based in an area famed for its crop circles, one of Honeystreet's brews is called 'Croppie', while another is known as '1810' – the year the Kennet and Avon Canal opened. Honeystreet Ales also own The Barge Inn in Honeystreet right beside the Kennet and Avon Canal; both the canal and pub opened around the same time, although back then the pub was known as The George. Fans of Inspector Morse might just recognize the pub as it was used in the filming of Colin Dexter's *The Wench is Dead*.

Ramsbury Brewery started brewing in 2004, following in the steps of a brewing heritage in the village that existed back in the 1790s, when Ramsbury was 'noted for the excellent beer of which there is a great consumption in London'. The farming estate which owns the brewery covers over 10,000 acres, including woodland and chalk grassland for sheep grazing; however, the predominant land use is the production of grain. The estate produces around 2,000 tons of malting barley – including sought-after varieties such as Maris Otter and Chariot – per year, much of it being exported to brewers all over Europe.

Much larger breweries exist just outside the AONB, with John Arkell having started the Arkell Brewery in Swindon in 1843 and Henry Alfred Wadworth founding Wadworth & Co Ltd in 1875 in Devizes.

The Lambourn Valley Cider Company was formed in 1995 to produce cider from a range of local apples; sadly in 2012 they took the decision to cease commercial production. Kintbury is home to Ciderniks which started out in 2003, using mainly local fruit grown in the Kintbury and Inkpen area, as well as importing some apples from Herefordshire; their Yellowleg cider takes its name from the old nickname for Inkpen clay workers.

Wyatt's Cider in Cold Ash started out in 2006 using apples from orchards to the east of Newbury, while Tutts Clump Cider near Bradfield uses a mixture of cooking, eating and crab apples as well as traditional cider apples. Originally started as a hobby in 2006, Tutts Clumps Cider went commercial in 2008.

At Upton, the Upton Cider Company produces cider using apples from their own orchard that was planted in 1970. Originally the apples were destined for the Taunton Cider Company; however, in 1983 they decided to start making their own cider on the farm.

The well-drained chalk soils that are found within the North Wessex Downs are ideal for growing certain varieties of grape and there are now a few small vineyards producing white, red and rosé as well as sparkling wines using the 'champagne method'. In the far northeast corner of the AONB, close to the River Thames, is the 14-acre Brightwell Vineyard where several varieties of grape are grown, including Reichensteiner, Bacchus, Huxelrebe, Chardonnay and Dornfelder. At the foot of the Berkshire Downs, along the northern edge at the picturesque village of East Hendred, is the 9-acre Hendred Vineyard which has been producing wines since the 1970s, from grape varieties such as Seyval Blanc, Madeleine Angevine and Pinot Noir.

To the south of Devizes, just outside the AONB, is the village of Littleton Panell and the a'Beckett's Vineyard. They planted their first 5,000 vines in 2001 with a further 4,300 being planted in 2010, and the vineyard now covers around 10 acres. Grape varieties for white wine include Chardonnay, Pinot Auxerrois, Reichensteiner and Seyval Blanc, and for red wine they include Pinot Noir and Dunkelfelder.

GALLOPING HORSES

Horse racing is big business in the North Wessex Downs. The area is second only to Newmarket and around 10 per cent of Britain's racehorse trainers are located within the AONB, with around 2,000 racehorses being trained there; a business that directly employs around 800 people, with many more supplying associated services. The main concentrations of training stables are to be found around Marlborough, East and West Ilsley, Kingsclere and, in particular, Lambourn. This last centre has such a wealth of horse racing tradition that it is known as the 'Valley of the Racehorse', complete with equine swimming pools, an equine hospital, six all-weather gallops, over 600 acres of downland turf training ground and around twenty-five training yards, as well as being the base for the National Trainers Federation.

The rolling chalk downs with their soft, springy turf, coupled with the free-draining nature of the soil that ensures the ground does not get too 'heavy' during the wetter winter months, makes chalk grassland ideal horse-training terrain. A quick look at an Ordnance Survey map reveals the large number of horse gallops (or training areas) associated with the training establishments that are located across the AONB.

Throughout history, horses have been bred for specific tasks, whether for agriculture, transport or war. However, it wasn't until the late seventeenth and early eighteenth centuries that the systematic breeding of horses was developed, leading to the introduction of the modern Thoroughbred horse. All modern Thoroughbred horses can trace their pedigree to just three imported stallions: Byerley Turk, Darley Arabian and Godolphin Arabian.

Modern horse racing is said to have originated around the time of the Stuart kings at the start of the seventeenth century, and James I established stables at Newmarket where he kept racehorses and 'riders for the

▲ The Lambourn Valley is known as the 'Valley of the Racehorse' due to its large number of horse training establishments

▲ Horse riders on the Ridgeway

races' – the first royal jockeys. However, it was during the reign of Charles II that horse racing – the 'sport of kings' – flourished and developed. The increasing interest in horse racing led to the development of racecourses throughout the country, with some on the North Wessex Downs.

In 1727 William Craven, 3rd Baron Craven (1700–1739), racehorse owner and breeder, started the Wantage Racehorse Meeting on land he owned near Letcombe and in 1731 he started the Lambourn Racehorse meetings on Bailey Hill Down near to his Ashdown House property; the meetings continued to be held there until the land was enclosed in 1803. His brother, Fulwar, 4th Baron Craven (1702–1764), founded both the Craven Hunt and the Ashdown hare coursing meetings in 1740. He also commissioned the artist James Seymour to produce a series of sporting paintings, including probably the most famous, *A Kill at Ashdown Park*, in 1743. The painting remained in the family for over two hundred years before being sold at auction in 1968; it now hangs in Tate Britain. Later, William, 6th Baron Craven (1738–1791), commissioned a series of paintings of his beloved racehorses by the artist Francis Sartorious (1734–1804).

At one time there were a number of racecourses within the AONB, at venues such as Lambourn, Letcombe, Hungerford, Bucklebury and Marlborough – the latter was in use from 1730 until 1873. However, the enclosure of common lands on which some of them were sited brought about a gradual decline in numbers. Although there are no professional racecourses within the AONB, there are a few point-to-point racing courses (steeple chase racing for amateur riders on horses that have been out hunting) including Lockinge, home to the Old Berks Point-to-Point which was first held in 1953, and Barbury Racecourse. Since the early 1900s racehorses have been trained on the gallops around Barbury Castle. Yet in 1962 the original Barbury Castle racecourse was turned over to arable farming and it was not until thirty years later that the current point-to-point racecourse was re-opened.

In 1805 another William Craven, this time the 7th Baron Craven, who was created the 1st Earl of Craven (second time round) by George III, started the annual two-day Newbury Races Meeting on his land at Enborne Heath. In 1811 the races moved to Woodhay Heath until the last meeting in 1815. Almost another ninety years passed before the present Newbury Racecourse (located outside the AONB) came into existence following a proposal from the Kingsclere-based trainer John Porter. The first race, held in 1904, was the Whatcombe Handicap. Racing at Newbury stopped during both world wars, the course being used as a prisoner-of-war camp during the First World War and then as a supply depot by the Americans during the Second World War. Racing resumed in 1949, which is now home to the Hennessy Gold Cup.

The 2nd Earl Craven (1809–1866) continued the sporting traditions of his forebear; his horse Charity won the Grand National in 1841 and another of his horses, Wild Dayrell, trained by Francis Popham, won the Derby in 1855. He also revived the Lambourn Racehorse Meeting on Weathercock Hill to the east of Ashdown House.

By the middle of the nineteenth century, the area around Lambourn was starting to gain favour as a great place to train horses. Charles William Jousiffe (1845–1891), jockey and racehorse trainer, came to live at Seven Barrows just outside Lambourn in 1875, with a couple of horses and a small stable. He came to be known as the 'father of horse racing in Lambourn'. The lychgate at Lambourn's medieval Church of St Michael was dedicated to his memory. Yet it wasn't until the time of the National Hunt trainers Fulke Walwyn and Fred Winter, who won a number of Cheltenham Gold Cups, Champion Hurdles and Grand Nationals, that Lambourn's reputation was finally established. Between them, Walwyn and Winter won over 3,500 races from their Saxon House and Uplands yards. In addition to his other successes, Walwyn trained the late Queen Mother's jumpers and in the 1960s six of the Queen's yearlings were sent to West Ilsley to be trained on the flat by the late Dick Hern.

In addition to its National Hunt successes, the area also has a strong history of successful flat race trainers, including Peter Walwyn (cousin of Fulke Walwyn), Barry Hills, Jamie Osborne, Brian Meehan and Marcus Tregoning – the last-named training at Sheikh Hamdan Al Maktoum's purpose-built Kingwood House Stables near Lambourn.

Many famous horses have been trained at Kingsclere, including Mill Reef. Born in the United States, the horse was trained by Ian Balding (father of TV presenter Clare Balding) at Park House Stables in Kingsclere and went on to win twelve out of fourteen races (including the Derby and the Prix de l'Arc de Triomphe) between 1970 and 1972. Park House Stables was, in fact, where John Porter, aforementioned founder of Newbury Racecourse and said to be the most successful trainer of the Victorian era, trained an impressive number of racehorses, including seven Derby winners.

IRON FOUNDRIES

The knowledge of how to make iron came to Britain around 700 BC and iron started to replace bronze as the material of choice for both tools and weapons. The earliest form of smelting, known as bloomery smelting, took place in a furnace, consisting of a pit with a clay chimney above in which a mix of iron ore and charcoal was burnt. In order to reach the high temperature required (around 1,200 degrees Celsius), bellows were used to blow air into the furnace. This process produced an iron bloom, which contained both voids and slag. The bloom was then worked in a forge where it was heated and hammered to drive most of the slag out and produce wrought iron. This could be shaped to make implements or bars that could be used by the blacksmith to create tools and weapons.

During the Middle Ages water power was often used both to drive the bellows and to hammer the iron bloom.

A more efficient form of furnace, known as a blast furnace, started to emerge in Britain around 1500. These furnaces operated at a higher temperature and were capable of producing liquid iron. Limestone was added to ore as a flux to increase the yield of iron by producing a calcium-rich, rather than an iron-rich, slag. These types of furnace would work continuously for weeks at a time, with the liquid iron being tapped off at intervals to be cast into numerous small ingots. These castings, known as pigs, were linked to a main channel along which the liquid metal flowed and were said to resemble a sow feeding piglets – hence the name pig iron (pig iron was later re-melted and worked in a forge).

Ramsbury certainly had a brass and iron foundry in the 1850s – although around a thousand years earlier there was a Saxon iron foundry located where the High Street is now. Bucklebury had an iron foundry with a waterwheel on the River Pang, owned by the Hedges family from the 1600s to 1904; the foundry closed in the 1950s. Compton had the Whitewall Ironworks run by Thomas Baker and Sons from the early 1800s; the works were still operating up until the 1950s.

Wootton Rivers had a couple of foundries; Thomas Holmes was working there in 1848, and between 1855 and 1859 a firm of iron founders and agricultural implement makers, Oatley & Morris (later Oatley & Whatley), started trading there before moving to Pewsey in the 1860s. The old Whatley's foundry off the High Street in Pewsey, started by George Whatley, was home to an iron and brass foundry during the nineteenth century; the building now houses The Heritage Centre. In Collingbourne Ducis the Bourne Iron Works were established by James Rawlings in the 1860s; the foundry closed at the start of the Second World War.

At one time Hungerford had two iron foundries: Gibbons' and Cottrell's. Richard Gibbons, along with his brother James, had begun work as an iron founder in Ramsbury in 1819, before moving to Hungerford in 1824 and opening a foundry in Bridge Street. Both Hungerford and Kintbury were a focus for agricultural 'Swing Riots' in 1830 and a considerable sum of damage was done to the iron works. The business finally closed in 1931. The other foundry – the Eddington Iron Works – was started by Levi Cottrell around 1870 and closed in 1911.

CHALK AND BRICKS

Whiting, sometimes known as whitening, was manufactured at several locations within the AONB including around Kintbury. In 1862 there were said to be five manufacturers of whiting in Kintbury, producing around 1,800 tons per year, which was transported by boat along the adjacent Kennet and Avon Canal; by 1905 only one manufacturer remained.

Whiting is made by grinding pure chalk (calcium carbonate) into a fine powder before washing it in water and allowing the particles to settle. The resultant sediment is then moulded into cakes or balls before being allowed to dry. Whiting has been used in a diverse range of applications from the 'bleaching' of ships' sails to products such as paint, polish, putty and even toothpaste.

Chalk was quarried to use as a conditioner to help improve the acidic soils as well as being added to clay when making bricks. Former chalk mines still exist between Yattendon and Burnt Hill, although all the entrances except one were closed off in the early 1900s. Exploration of the tunnels has revealed graffiti scratched onto the mine walls, dated 1720.

Chalk is a rather soft and porous rock which makes it less than ideal as a building material. However, chalk stone, known locally as 'clunch', has been quarried in the past and used in old cottages in villages such as Ashbury and Woolstone along the northern reaches of the North Wessex Downs. The cottages have resisted the weather for centuries by having what is often known as 'good shoes and a hat', in other words the cottages are built on stone or brick to stop rising damp, while on top is an overhanging roof of thatch to resist the weather.

The Romans were the first people to bring hard-fired (as opposed to air-dried) brick-making to the area, yet after they left Britain, brick-making died out, and buildings made in the Dark Ages used bricks removed from older buildings. Brick-making restarted in the fifteenth and sixteenth centuries; however, at this time they were expensive to produce and were only used for important buildings, the bricks being made on an individual basis for each building. During the reign of Elizabeth I, bricks that we recognize as Tudor became standard, being known as 'statute' bricks, their size of 9 x 4½ x 2¼ inches being set by law in 1571 (for comparison, modern metric bricks are 215 x 102.5 x 65 mm, or approximately 8½ x 4 x 2½ inches). To make such bricks clay was dug out in the winter and allowed to weather before being mixed with other ingredients such as chalk or sand in a

▲ The Pot Kiln pub at Frilsham, showing classic Flemish Bond brickwork; the bricks were made at the old kilns that were once located here

pug mill (from the mid-nineteenth century, horse-driven pug mills were used). The bricks were then moulded during the middle part of the year – an Act of Parliament only allowed this to occur from March to October as the quality of bricks made in winter was poor. Once moulded, the bricks were left out to dry for several weeks before being fired in a kiln.

One of the oldest methods of firing bricks was using a temporary structure or 'clamp' constructed from tapering layers of unfired, or 'green', bricks with fire tunnels left open at the base where the fires were lit, and chimney gaps rising up through the layers to allow heat to move through the bricks. The whole structure was dismantled after firing. Updraft kilns, known as Scotch or Suffolk kilns, were built from fired bricks, and flues ran under the perforated floor. Unfired bricks were stacked inside the chamber with small gaps between them to allow the heat to circulate and the open top was covered over to conserve the heat and aid even firing. The firing process took three or four days and the kilns had to be stoked regularly throughout the day and night.

A much more efficient way of firing bricks was the downdraft kiln, often called a circular or 'beehive' kiln. The heat from the fires was directed upwards by baffles and then drawn down inside the chamber from the domed roof, passing through the stacked bricks and then out through the perforated floor, being drawn through the kiln by the draught caused by the chimney.

Brick-making within the North Wessex Downs was limited in scale because of the limited supply of the raw materials, particularly clay. To the north of the region is an area with Oxford clays, then along the northern edge of the AONB is a band of Gault Clay. However, it is in the area through the centre of the AONB following the Kennet Valley east towards Newbury where we find the clays of the Reading Formation and London Clay Formation that were used by many of the local brickworks, such as the Dodsdown Brick and Tile Works at Wilton near Great Bedwyn. These produced a buff-coloured brick known as the Ailesbury brick – some of which have been used as detailing in cottages along the west side of Church Street near the church in Great

Bedwyn. Other brickworks in the Kennet Valley were located around Hungerford and Kintbury. However, owing to the limited supply of clay the local brick-making industry started to die out in the 1920s, moving to areas with larger amounts of clay, thus making the process much more economical.

One business that closed for a rather different reason was the brick and tile works beside the Pot Kiln pub in Frilsham that used locally occurring deposits of clay; the former kilns were located in the present car park. The brickworks were operated by the Barr family for many generations until 1940 when the 'black-out' regulations brought in during the war meant that the kiln had to be closed down; the kilns were open at the top and authorities were worried that German bombers could use them as a night-time navigation aid.

During the nineteenth century the Pinewood Brick and Tile Works were established at Hermitage to the west of Frilsham. The site, which closed in 1967, had its own railway siding connected to the former Didcot, Newbury & Southampton Railway. Further west there were brickworks at Curridge and Wickham.

The way bricks are laid when making a wall is called the bond and there are several common types that may be seen when looking at brick buildings throughout the North Wessex Downs. English bond was one of the first to be used and was common throughout the sixteenth and seventeenth centuries, using differing layers of stretchers (side-on bricks) and headers (end-on bricks), with the headers centred on the stretchers and each alternate row vertically aligned. Flemish bond became popular in the eighteenth century, using layers of alternating stretcher and header bricks; this was a more decorative design and often used with alternate-coloured bricks. Other forms of bricklaying included stretcher bonds, where layers of side-on bricks are used, and header bonds using layers of end-on bricks.

WIND AND WATER FOR POWER AND FOOD

Rivers throughout the North Wessex Downs have been used as a source of power and food from well before the time of the Domesday Book, from fishing to grinding grain, fulling woollen cloth and manufacturing paper.

The Domesday Book records that there were water-driven mills along the River Pang at Yattendon, Frilsham, Bucklebury, Stanford Dingley, Bradfield and Pangbourne.

Windmills and Water Mills

The mill at Tidmarsh, a corn mill driven by water from the River Pang, was functioning in the early thirteenth century and was, for the most part, held by the Abbot of Reading until the Dissolution of the Monasteries. The old mill probably stood on the site of the present Mill House; more recently the house became a retreat for members of the Bloomsbury Set in the early twentieth century (see page 145). Tidmarsh Manor also had a fulling mill in the late sixteenth century, though this has long since disappeared.

For centuries, man has used wool to produce cloth; stone whorls that would have been attached to the weaving spindles have been found dating back to the Iron Age. Yet it was not until the twelfth century that cloth-making started to change rapidly with the invention of the horizontal loom that could make much larger lengths of cloth.

After weaving, woollen cloth needs to be cleaned of oils and made thicker in a process known as fulling (this consisted of stamping or walking on the cloth in a trough of water combined with a cleaning agent); from the medieval period this was usually carried out in a water-powered fulling mill where the cloth was beaten by large wooden hammers. The final stage of the process was to stretch the cloth on large frames known as 'tenters' with the cloth being held by metal hooks called 'tenterhooks' to dry out – it was from this process that the phrase 'on tenterhooks' is derived, meaning to be held in a state of suspense.

Many of the towns and villages within the North Wessex Downs were involved in the cloth industry, including Marlborough and Hungerford, while the rolling chalk grassland of the downs was used to rear vast numbers of sheep.

The importance of the cloth industry grew in the Middle Ages, reaching its height in the Tudor period when the cloth industry dominated English exports; one

▲ The former Denford Mill on the River Kennet near Hungerford

▲ Wilton Windmill – the only remaining windmill in the North Wessex Downs

of the more important centres close to the North Wessex Downs was Newbury, home to the famed clothier John Winchcombe (d.1519), aka Jack of Newbury.

Some of the former water-powered mills, such as the one at Kintbury close to the Kennet and Avon Canal, and the old fulling mill at Denford Mill, just east of Hungerford, have survived, having been converted into private residences. The former water-powered corn mill at Bagnor (the first mention of a mill at Bagnor was in the Domesday Book) was converted into The Watermill Theatre in the early 1960s.

Several paper mills were located along the Kennet Valley at Sheffield Bottom, Sulhampstead, Thatcham, Newbury and Bagnor, all of which were helped by the opening of the Kennet Navigation in 1723.

A rare find is Wilton Windmill, the only working windmill within the North Wessex Downs. The mill was built in 1821 to replace a former water mill that ceased operating when the water was diverted from the River Dun to the Kennet and Avon Canal. Standing on an exposed chalk ridge at 550 feet, the mill was operational for a hundred years until the advent of more efficient forms of power, particularly electricity, brought about its demise. After standing derelict for fifty years the mill was restored and today produces stone-ground flour which is for sale on site and at local outlets. Built from brick, the mill has a fantail that keeps the sails aligned with the wind, acting as an automatic rudder.

Other places within the AONB had windmills, though little remains of them today. There was once a post mill at Compton, located to the north of the church. The mill was originally located at Little Hungerford near Hermitage before being moved to Compton around 1760. Records suggest that it was built in 1742, though by 1900 it was already in a ruinous state.

Food from the Rivers

The crystal-clear waters of chalk streams and rivers are well known for their trout fishing; however, they were also once famed for their watercress. Known since

▲ The remains of the watercress beds on the Letcombe Brook at Letcombe Bassett

ancient times for its health-giving properties, watercress is a semi-aquatic indigenous plant with a peppery taste, packed with nutrients and vitamins. One of the main requirements for growing watercress is a good supply of clean, mineral-rich, well-oxygenated water at a fairly constant temperature and the chalk streams of the North Wessex Downs fit the bill perfectly.

In the eighteenth and early nineteenth centuries watercress sandwiches were a staple part of the working-class diet as it could easily be picked from rivers and streams. It was often known as 'poor man's bread' as many impoverished labourers had access to watercress but not the bread, so it was eaten just on its own. Although the plant was used for many centuries, the first watercress farms did not appear until the early 1800s.

Watercress was commercially grown at the Kimberhead Spring and the Blue Pool near Stanford Dingley. Here, the clear spring water from the artesian aquifer bubbles to the surface forming the perennial head of the River Pang; watercress production here ceased in the 1980s.

Letcombe Bassett and the Letcombe Brook were once famed for their watercress, and 'Bassett cress' was a familiar cry in London's Covent Garden Market. All that remains of this once-flourishing trade are the ruins of the old gravel beds in which the watercress was grown. Even Thomas Hardy referred to the village as *Cresscombe* in his novel *Jude the Obscure*.

Watercress was also grown on the Shalbourne Stream, a tributary of the River Dun, which flows past the Wiltshire village of Shalbourne, from the 1920s until 1972. Varieties such as 'morning dew' and 'mill brand' were sent to both Birmingham and Covent Garden in London. On the River Kennet around Ramsbury and neighbouring Axford, watercress was grown from the 1860s until the 1980s.

Down at St Mary Bourne in Hampshire, watercress is still commercially farmed on the headwaters of the Bourne Rivulet. The farm was developed by Eliza Fleet

– also known as Eliza James – who at one time was said to be the biggest owner of watercress farms in the world. For fifty years Eliza was a well-known figure at Covent Garden Market; having devoted much of her life to watercress she was known as 'the watercress queen'; at her funeral a huge wreath of watercress was placed on her coffin. At the time of her death, her company, James & Son, who held the trademark Vitacress, was selling fifty tons of watercress every week; Vitacress Salads still operate the farm at St Mary Bourne.

CUTTING EDGE SCIENCE AND INNOVATION

The former Second World War airfield at Harwell, close to the northern edge of the North Wessex Downs, recently rebranded as Harwell Oxford (one of the UK's two National Science and Innovation Campuses), became the home to Britain's first Atomic Energy Research Establishment (AERE) when the Ministry of Supply took over the site in 1946. This later became part of the newly formed United Kingdom Atomic Energy Authority (UKAEA) in 1954.

At one time there were five research reactors, including the first nuclear reactor in Western Europe known as GLEEP – the Graphite Low Energy Experimental Pile – which operated for forty-three years. BEPO (British Experimental Pile '0') was the second UK reactor, commissioned in 1948 and used to demonstrate the viability of commercial power reactors; this reactor was a forerunner of the Windscale Piles reactor.

BEPO was closed in 1968, when it was superseded by the Material Testing Reactors, DIDO and PLUTO, that were first commissioned in 1956 and 1957. These reactors were mainly used for fuel and material testing – materials tested for several months in these reactors demonstrated the effects of irradiation over a period of twenty to thirty years in a civil nuclear reactor. All of these reactors have now been closed and the long-term plan is to remove all the radioactive waste from the site.

However, cutting-edge science has continued at Harwell and the site is home to the Rutherford Appleton Laboratory (named after the physicists Ernest Rutherford and Edward Appleton) which includes the ISIS neutron source and the Diamond Light Source Synchrotron. The ISIS neutron source – the world's leading pulsed neutron facility – produces beams of neutrons and muons (a neutron is one of the three basic components of the atom, the other two being the proton and the electron; a muon is an elementary particle similar to an electron) that allow scientists to study materials at the atomic level using a suite of instruments, often described as 'super-microscopes'.

The Diamond Light Source Synchrotron – the large silver-coloured metal disc-shaped building visible from the Ridgeway – is the largest UK-funded scientific facility to be built for over forty years. Here electrons are accelerated to near light-speed, generating brilliant beams of 'light' – known as 'beamlines' – from infra-red to X-rays. These beams of light are used by researchers to conduct experiments in a wide range of disciplines from health and medicine to solid-state physics and engineering. Opened in 2007, with an investment in excess of £400 million, the facility now offers eighteen 'beamlines' although further investment in the facility will bring the total number of 'beamlines' to thirty-two by 2017.

The scientific community at Harwell Oxford continues to grow as more investment is made in the facilities; recent openings include the Satellite

▲ A canal boat on the Kennet and Avon Canal adds a splash of colour

▲ Canal boat passing through Hungerford Lock on the Kennet and Avon Canal

Applications Catapult (catapults, formerly known as Technology and Innovation Centres, are centres of excellence that bridge the gap between business, academia, research and government).

Boats, Trains and Cars

Sparsely populated areas within the North Wessex Downs have little in the way of transport links save for a few narrow country lanes and farm tracks. However, the area has its fair share of major transport links from the prehistoric days of the Ridgeway to the turnpike roads of the eighteenth century and the now ever-busy M4 motorway, and from the leisurely pace of canal boats on the Kennet and Avon Canal to the great railways built by the Victorians.

THE KENNET AND AVON CANAL

The Kennet and Avon Canal, which celebrated its bicentenary in 2010, was formed by the joining of two waterways in the early nineteenth century. The Kennet Navigation from the River Thames at Reading to Newbury, which had opened in 1723, and the Avon Navigation from Bath to Bristol, which opened in 1727, were joined by the construction of a 57-mile-long canal, giving a total canal length of 87 miles.

The bill to make the River Kennet navigable from Reading to Newbury was introduced into Parliament in 1708. Despite protests from traders in Reading, who feared a loss of custom when people no longer had to travel to the town to buy and sell goods, the Kennet Navigation Bill received Royal Assent in 1715.

By 1770 people were starting to put forward the idea of building a new canal that would link the River Kennet and the River Avon, yet it wasn't until 1790 that a decision was made to proceed with plans for the proposed canal taking a route from Hungerford via Ramsbury and Marlborough. However, because of growing fears about the availability of a good supply of water for the original route, a more southerly route, favoured by John Rennie, was proposed in 1793. The new route would pass through Great Bedwyn and the Vale of Pewsey, with an extension canal to Marlborough. The drawback with this route was the need for a 2½-mile-long summit tunnel to be constructed between Crofton and Burbage.

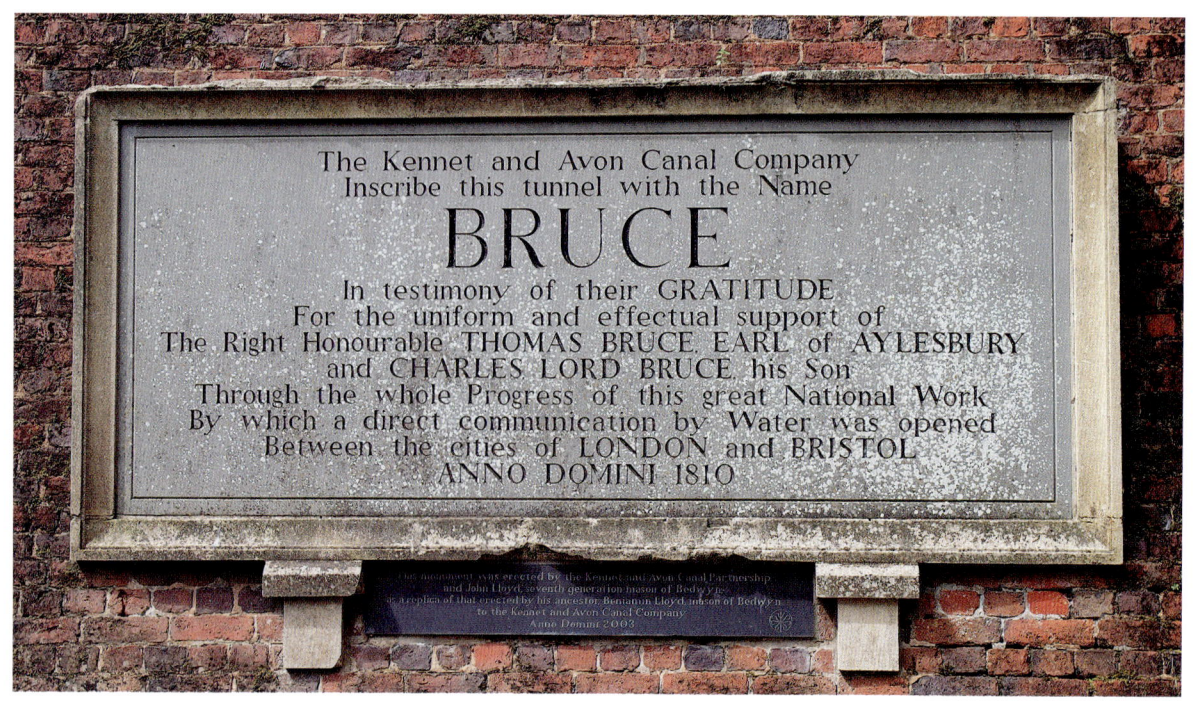

▲ Sign commemorating the Bruce Tunnel on the Kennet and Avon Canal

Plans for the extension to Marlborough were shelved and in 1794 the proposal received Royal Assent, with John Rennie being appointed chief engineer. A further survey recommended altering the route yet again, this time raising the summit by about 40 feet to reduce the length of the tunnel – known as the Bruce Tunnel – to approximately 500 yards, thereby substantially decreasing the cost and time required to construct the canal, even though more locks and a steam-driven pump would be required to raise water to the new canal summit. The tunnel, named after Thomas Brudenell-Bruce, Earl of Ailesbury, and his son Charles Brudenell-Bruce, who owned the Savernake estate through which the canal was passing, had no towpath, which meant that the horse-drawn boats had to be pulled through the tunnel by the boatmen using chains fixed to the walls, while the horses were taken over the top.

Finally, in October 1794, work started on the canal, although the financial impact of the Napoleonic Wars resulted in the build schedule being extended. The section from Newbury to Kintbury opened in 1797, followed by Hungerford in 1798 and Great Bedwyn in 1799. The section between Great Bedwyn and Devizes took a further ten years to build and in 1810 the Kennet and Avon Canal was finally open, giving a direct trade route between London and Bristol.

The canal forms a distinctive linear feature, threading through the heart of the North Wessex Downs, following the Kennet Valley to Hungerford before heading south-west through Great Bedwyn, where it follows the River Dun, and then west through the Vale of Pewsey before leaving the AONB to pass Devizes and descend a magnificent flight of twenty-nine locks at Caen Hill to overcome a height difference of 237 feet in a short 2-mile section with a 1 in 44 gradient.

Located near the summit of the canal, beside Wilton Water, is the world-famous Crofton Pumping Station. Built in 1807, the pumps were used to raise water 40 feet from natural springs at Wilton up to the summit of the canal, using a specially built 'leat' or feeder culvert, thereby replenishing the water lost each time a boat went through a lock. The first engine installed was a 36-inch-bore Boulton and Watt and in 1812 a second 42-inch Boulton and Watt engine was also installed,

▲ The Crofton Pumping Station houses the world's oldest working steam-driven water pump

although in 1846 the original engine was replaced with one made by Harvey of Hale. The remaining two beam engines were capable of raising about 1 ton of water with each single stroke. In response to increasing traffic along the canal, Wilton Water was created in 1836 to provide a larger store of water that could be pumped into the canal, rather than relying on what was available from the natural springs.

Along the length of the canal there were a number of wharfs built at places such as Hungerford, Great Bedwyn, Burbage and Pewsey, so that goods transported along the canal including coal, timber and grain could be easily loaded and unloaded. For forty years the canal prospered, but the passing of the Great Western Railway Act in 1835 led to the building of Isambard Kingdom Brunel's railway, which offered a faster and more efficient transport route between London and Bristol. In 1852, the Great Western Railway Company succeeded in buying the Kennet and Avon Canal, though on the proviso that it was kept open as a canal. However, they made little attempt to maintain the route over the next hundred years, leading to its gradual decline; they even offered traders preferential tolls to use the railway instead.

During the Second World War, with the threat of a German invasion, the near-derelict canal became part of Britain's defences when, as mentioned in Chapter 2, it was used as the GHQ Stop Line Blue.

By 1955 the canal was in very poor condition, though plans to abandon it completely were thwarted by public opposition. The Kennet and Avon Canal Trust purchased the Crofton Pumping Station from British Waterways in 1968, with the objective of restoring it to full working order, and two years later the steam engines were working again. Although electric pumps have been installed to pump water into the canal, these magnificent Cornish Beam engines are still used on several occasions throughout the year; the 1812 Boulton and Watt is the oldest working beam engine in the world.

The full length of the canal, restored and cared for by the Kennet and Avon Canal Trust, was formally re-opened by the Queen in 1990. Nowadays the canal forms a popular heritage tourism destination for boating, walking and cycling as well as being an important feature for wildlife conservation.

RAILWAYS PAST AND PRESENT

The mid-nineteenth century witnessed the arrival of the railways, which reached their peak during the early part of the twentieth century before the increasing use of road transport brought about the closure of many lines.

Lines That Survived

Isambard Kingdom Brunel's famous Great Western Railway (GWR), originally known as the 'Bristol and London Railroad', was constructed to connect London with Bristol in the West Country. The line passes through Pangbourne in the north-eastern corner of the AONB en route from Reading to Didcot and then on to Swindon; the section to Swindon opened in late 1840 and the line to Bristol was fully open by the summer of 1841. The original plan was to run the railway through Savernake Forest near Marlborough, but the Marquess of Ailesbury, who owned the land, objected and a more northerly route through Swindon was chosen.

Originally the railway was constructed using Brunel's wider 'broad' gauge (7 feet ¼ inch) track; this gave better passenger comfort but made construction more expensive, and caused difficulties when it had to interconnect with other railways using the narrower gauge. As a result of the Railway Regulation (Gauge) Act 1846 the gauge was changed to standard gauge (4 feet 8½ inch) throughout the GWR network, although this took some forty years to implement fully across their network.

With the line passing through Swindon, a decision was made to build a service and repair works there in 1841; by 1846 the Swindon factory was producing new trains. In its heyday, the Swindon Works employed over 14,000 people, before a gradual decline saw the works finally close in 1986. The last steam locomotive to be made in Britain – the 92220 *Evening Star* – was built here in 1960; one of the old buildings now houses STEAM – a museum dedicated to the Great Western Railway, including the GWR express passenger locomotive, 4073 *Caerphilly Castle*.

The GWR-backed Berks & Hants Railway, also designed by Isambard Kingdom Brunel, comprised two lines: a 14-mile branch line from Reading to Basingstoke, which opened in 1848, and a 25-mile line from Reading to Hungerford, which opened a year earlier in 1847 and ran parallel to the Kennet and Avon Canal, passing through Newbury and Kintbury. In 1862 this line was extended westwards via Great Bedwyn and Pewsey to link with the Wilts, Somerset & Weymouth Railway. Originally built as a broad gauge line, it was converted to standard gauge in 1874. This now forms part of the line from Reading to Westbury; the branch line serving Devizes closed in 1966.

The line to Basingstoke was extended to Andover in 1854, passing Overton and Whitchurch and crossing the Bourne Rivulet via an impressive brick viaduct located between St Mary Bourne and Hurstbourne Priors. There was a station here, known as Hurstbourne Priors, but this closed in 1964. The line finally reached Exeter in 1860. This line now forms the West of England Main Line running from London Waterloo to Exeter, passing along the lower edge of the AONB. From Hurstbourne Priors the London & South Western Railway (LSWR) built a short branch line to Fullerton, just south of Andover, where it connected with the now-closed line from Andover heading towards Southampton. Passenger services were withdrawn in 1931 and, although the branch line stayed open for freight traffic, it finally closed in the 1950s.

Lines That Weren't so Lucky

The 12-mile long Lambourn Valley Railway (LVR) between Newbury and Lambourn opened in 1898, passing through Speen, Stockcross, Boxford, Welford, Great Shefford, East Garston and Eastbury. Originally operated as a privately owned railway, the line merged with Great Western Railways in 1905. The majority of the line closed to passengers in January 1960, though the section between RAF Welford and Newbury operated until November 1973; the final closure was brought about by the opening of the M4 motorway and the ending of an agreement between RAF Welford and British Railways, allowing the switch from rail to road for transporting supplies to the RAF site (now used as an ammunition store). Parts of the 19½-mile Lambourn Valley Way that runs from the Uffington White Horse

▲ The former rail station at Compton (now a private house)

high up on the downs to Newbury follows sections of this old line.

The 45-mile north-south Didcot, Newbury & Southampton Railway (D, N & S) was built in the nineteenth century to connect Didcot with Southampton. The northern section between Didcot and Newbury, including the station at Compton, was opened in 1881. Compton became an important centre for the transportation of sheep to and from the famous sheep fair at East Ilsley. The line was absorbed by Great Western Railway in 1923 and the section between Didcot and Newbury closed in 1962.

Most of the northern section, which had stations at Upton, Churn, Compton, Hampstead Norreys, Pinewood Halt and Hermitage, can still be seen; and the old station at Compton still exists (now a private house, although a public footpath passes right by the former station). To the south of Newbury, much of the A34 now follows the old railway that had stops at Woodhay, Highclere and Burghclere before heading south of the AONB via Whitchurch.

The northern section of the Swindon, Marlborough & Andover Railway (SMAR) from Swindon to Marlborough opened in 1881, with the southern section from Grafton to Andover opening a year later (in 1884 the Swindon, Marlborough & Andover Railway became part of the Midland & South West Junction Railway). For the section between Marlborough and Grafton, trains had to run on the existing GWR Marlborough branch line and part of the GWR Berks & Hants Extension Line until SMAR completed the missing link with the opening of the Marlborough & Grafton Railway in 1898.

Marlborough at one time had two separate railway stations, built by competing railway companies, although neither of them has survived today. As well as the Swindon, Marlborough & Andover line, the Berks & Hants Extension railway built a short branch line from Hungerford to Marlborough in 1864; however, this closed in 1961.

In 1930 an extra stop was added, known as Chiseldon Camp Halt, to service the nearby army base. However, following the end of the Second World War, traffic started to fall dramatically and the line was finally closed in 1961. Today, the old track bed between Chiseldon and Marlborough forms the 7½-mile Chiseldon and Marlborough Railway Path which, in turn, is part of the National Cycle Network (Route 482).

The boundary of the North Wessex Downs in the north-eastern corner follows part of the Cholsey & Wallingford Railway. This was originally to have been a branch line from Wallingford to Watlington connecting with the existing Great Western Railway that ran between Pangbourne and Didcot. The first part of the line from Cholsey to Wallingford, known locally as 'the bunk', opened in 1866, but because of poor traffic the planned extension to Watlington was never built. The line was sold to GWR in 1872 and closed to passengers in 1959; freight services to an industrial site just outside Wallingford ceased in 1981. At this point, the Cholsey and Wallingford Railway Preservation Society was formed to revive the surviving section of line as a heritage railway; Wallingford's railway station had already been demolished and the section of the line through the town had been built on. Trains started running on the preserved line in 1985 and both steam- and diesel-hauled trains operate at certain times throughout the year.

One That Never Made It Off the Drawing Board

The Light Railways Act 1896 was implemented to facilitate the construction of railways in rural areas; before this all new railways required a specific Act of Parliament before they could be built, greatly increasing the cost. The Light Railways Act limited both the speed (to 25 mph) and the weight of the train (8 tons per axle) allowing a lower specification for both track and bridges, with the aim of reducing the overall cost.

Following the introduction of the act, a number of lines were built, including the Basingstoke to Alton Light Railway (to the south of the AONB), which opened in 1901, and in 1900 a prospectus was issued for the Highclere, Kingsclere and Basingstoke Light Railway. The line was intended to connect with the Didcot, Newbury and Southampton Railway at Burghclere and pass through Ecchinswell, Kingsclere, Wolverton and Ramsdell en route to Basingstoke. Despite public support, the line was never built.

ROADS, OLD AND NEW

The North Wessex Downs region is home to some of the oldest 'green' roads in Britain and they don't come much older than those following the crest of the Marlborough and Berkshire Downs. These tracks, some of which now form part of the Ridgeway National Trail, were part of an ancient network of routes that stretched across southern England and may have been in use for over 5,000 years. Prehistoric man travelled over these downs to trade goods including stone axes from the Lake District and clay pots from Cornwall. They also left behind many historical sites that these ancient trackways joined together, including the stone circle at Avebury, the long barrow at Wayland's Smithy, the famous Uffington White Horse, and several Iron Age hill forts.

In the Dark Ages invading Saxons and Vikings most likely travelled along this network of tracks. It was the Anglo-Saxons who gave us the first written evidence of these ancient routes calling them *'hyrcweg'*, which loosely translates as 'ridgeway' or tracks that followed areas of higher ground.

During medieval times, invaders were replaced by drovers driving their livestock to market. However, it was not until the Enclosure Acts of 1750 that the Ridgeway we see today was formed from the multitude of tracks that travelled along the crest of the downs, where previously travellers chose the driest or most convenient route. The Enclosure Act brought about a single route of between 40 to 60 feet in width, defined through the building of earth banks and the planting of hedges to prevent livestock straying into cultivated fields.

During their 400-year rule in Britain, the Romans built an expanding network of roads linking their main settlements. These roads typically followed fairly straight routes except where they had to negotiate natural obstacles such as a steep valley or hill. Most roads were built on a raised earth bank, or agger, with a metalled surface on top using local materials such as chalk, flint and gravel. The use of a cambered or sloping surface and parallel ditch helped with water run-off and drainage, making the roads usable throughout the year. Following the Roman withdrawal from Britain, their road network continued to be used for many centuries. Even today, parts of our modern road network make use of the routes that were originally built by the Romans.

The Roman roads that are still used within the North Wessex Downs include the following:

The Ermin Way ran north-west from *Calleva Atrebatum* (Silchester) via *Corinium Dobunnorum* (Cirencester) to *Glevum* (now Gloucester); parts of the route are now followed by the B4000 from Wickham to Lambourn Woodlands and then minor roads through Baydon and Wanborough.

The Portway headed south-west from *Calleva Atrebatum* to *Sorviodunum* (Old Sarum); parts are now followed by minor roads around Hannington and from St Mary Bourne to Andover.

A road ran east–west from *Londinium* (London) to *Aquae Sulis* (Bath) via *Calleva Atrebatum*, *Cunetio* (near Mildenhall) and *Verlicio* (about 2½ miles south-west of Calne); the section around Silbury Hill is followed by the present A4 and a byway follows the route over Morgan's Hill.

From *Cunetio* a road headed north passing close to Ogbourne St George on its way to *Durocornovium* (near Swindon); parts of this are followed northwards by minor roads from Mildenhall and then by the A346 from Ogbourne St George towards Swindon.

Another road from *Cunetio* headed south-east through Savernake Forest to *Venta Bulgarum* (Winchester); several minor roads now follow this route near Wilton and especially to the south of Tidcombe where it is known as the Chute Causeway as it makes a sweeping detour round the dry valley of Hippenscombe Bottom before continuing south-eastwards along Hungerford Lane, passing just to the east of Tangley.

Moving forward several hundred years, a survey in 1228 mentions 'The King's Street' running between Hungerford and Marlborough and passing through Savernake Forest; this later became part of the present-day A4.

Roads developed to meet local community needs and in 1555 an Act of Parliament made local parishes responsible for the upkeep of roads. Along many of these routes, coaching inns were built in the towns and villages, such as the Bear Inn at Hungerford – one of England's oldest coaching inns, first mentioned in 1494 – and the Waggon & Horses Inn at Beckhampton, built in 1669, which was mentioned by Charles Dickens in *The Pickwick Papers*. This was also the time of the highwayman, making travel a dangerous pastime.

In 1635, Thomas Witherings, whom Charles I had appointed as the Postmaster of Foreign Mails in 1631, put forward a proposal for a letter delivery service that later became the Royal Mail. To help with the delivery of the post, Charles I charged him with building six 'Great Roads' and one of these was the Great West Road – more commonly known as the Bath Road – between London and Bristol, the forerunner of today's A4, although the route has changed a bit over the years. The other 'Great Roads' radiated out to Dover, Edinburgh (via York), Holyhead (Chester), Norwich (Great Yarmouth) and Plymouth.

Later in the seventeenth century signs were being used at some junctions and more detailed maps were being drawn so travellers could find their way, and John Ogilby produced the first 'road atlas'.

The roads of the day were in a very poor state, deeply rutted and dusty in summer and a mud-bath in winter, but the solution was at hand with the introduction of Turnpike Trusts in the eighteenth century. Groups of wealthy landowners and businessmen agreed to finance the improvement and resurfacing of roads in return for the right to charge each user a toll; the name 'turnpike' comes from the spiked barrier used at the toll gate. From 1767, milestones became compulsory on all turnpikes,

▲ The thatched Waggon & Horses near Beckhampton

both to inform travellers of direction and distances and to help stagecoaches keep to a schedule. Reminders of these turnpike roads survive today; milestones can be seen along many of our main roads – such as the A4 around Hungerford and Marlborough – while at Tidmarsh, on the A340, the former octagonal toll house still survives; following the closure of the trust in 1872 the tollhouse became a private residence.

Along the east side of the High Street in East Ilsley is an old milestone dated 1776, a reminder of the former Oxford to Newbury turnpike which later became part of the A34. The milestone mentions that it is 9 miles to Newbury, 11 to Abingdon and 17 to Oxford. The nearby Swan Inn, a former coaching inn, was built to cater for people travelling along the north–south route.

The early part of the nineteenth century saw an increase in the number of stagecoaches operating between London and Bristol; in 1836 there were ten stagecoaches a day travelling along the Bath Road through Hungerford. However, the arrival of the

▲ An old milestone on the A4

▲ The North Wessex Downs has many great places to walk – here looking towards Cottington Hill near Watership Down

railway in the 1840s brought an end to stagecoach travel, many turnpike trusts were wound up, and in 1889 the newly formed County Councils were given responsibility for looking after the main road network, from which numerous winding rural lanes spread throughout most parts of the North Wessex Downs.

The two main arteries of a wider network of main roads passing through the AONB are the M4 motorway from London to Wales and the A34 which connects Southampton to Oxford and the M40. First proposed in the 1930s, the Ministry of Transport finally announced plans to construct the M4 in 1956, yet the section through the North Wessex Downs was not opened until late 1971. The A34 was built in 1922 between Winchester and Oxford, but was extended to Manchester in 1935. More recently, in the late 1990s, the construction of the Newbury Bypass led to some of the largest anti-road protests ever seen in Britain; these at the time were dubbed the 'Third Battle of Newbury', the first two battles having occurred during the Civil War.

Tourism

The North Wessex Downs has a great deal to offer visitors, and tourism is widely recognized as an increasingly important sector, one that the AONB authority is keen to develop; the revenue generated through tourism is a significant contributor to the local economy.

As mentioned earlier, there is a wealth of archaeological and historic sites that attract large numbers of visitors; the Avebury World Heritage Site receives in the region of 400,000 visitors a year, while the Ridgeway National Trail attracts in the region of 100,000 visitors a year. Additionally, there are numerous smaller, but no less important, sites, historic houses and other tourist attractions within the AONB.

Throughout the North Wessex Downs there is a vast network of public rights of way with footpaths (walkers only – yellow arrow), bridleways (walkers, cyclists and horse riders – blue arrow), restricted byways (walkers, cyclists, horse riders and carriage drivers – purple arrow), and byways (same as for a restricted byway plus

THE NORTH WESSEX DOWNS

motorcycles and motorized vehicles – red arrow). This network also includes parts of several long-distance routes:

Brenda Parker Way – a 27-mile section of the 78-mile-long path between Andover and Aldershot, developed by the North Hampshire Ramblers Group in memory of Brenda Parker. It passes through parts of the AONB visiting Burghclere, Highclere and St Mary Bourne.

Kennet and Avon Canal – the towpath follows the entire length of the canal from its junction with the River Thames at Reading to the River Avon, and 30 miles of the route heads east to west through the heart of the AONB.

Ridgeway National Trail – a 42¼-mile section of the ancient Ridgeway – probably the oldest green road in Britain, which is also open to cyclists and horse riders, heads through the northern part of the North Wessex Downs from near Avebury to cross the River Thames at Streatley and Goring.

▲ The start (or end) of the Ridgeway at Overton Hill beside the Bronze Age barrows – this national trail takes you on a journey through 5,000 years of history

▲ View across the Kennet Valley from the Wayfarer's Way near Ashmansworth

Thames Path National Trail – a short (8½-mile) section of the 184-mile trail which follows the River Thames from the Thames Barrier at Greenwich to its source in the Cotswolds, meanders along the eastern edge of the AONB from Streatley upstream towards Wallingford.

Test Way – a 13½-mile stretch of the 44-mile-long walking route that starts high on the chalk downs at Inkpen and follows much of the course of the River Test to Eling, where it joins Southampton Water.

Wayfarer's Way – a 16¾-mile section of the 70-mile trail which, like the Test Way, starts at Inkpen Beacon and heads south through Hampshire to the coast at Portsmouth.

White Horse Trail – part of this 90-mile trail, which meanders its way past all the white horses in Wiltshire, passes through the western reaches of the AONB, visiting seven of the eight white horses that lie within the North Wessex Downs (namely: Alton Barnes, Broad Town, Cherhill, Devizes, Hackpen Hill, Marlborough and Pewsey).

Throughout the region these rights of way give access to some spectacular scenery, picture-postcard villages and lovely stretches of isolated countryside that combine to form the primary attractions for both recreation and tourism.

In Savernake Forest, just to the south-east of Marlborough, Lord Cardigan has given access to a network of paths and cycle routes that allow visitors to enjoy the beauty of the ancient woodland and wood pastures that make up the Savernake Site of Special Scientific Interest (SSSI).

The Downs offer numerous viewpoints, including: Walkers Hill (SU109636) looking south across the Vale of Pewsey, Cherhill Down (SU047693) looking north and west across Wiltshire, Whitehorse Hill (SU300863) with views to the north across the Vale of White Horse, Combe Gibbet (SU364622) looking north across the Kennet Valley, Beacon Hill (SU457572) offering a 360 degree panoramic view of the North Hampshire Downs, and Watership Down (SU496568) looking north across the Kennet Valley.

▲ Canal boat on the Kennet and Avon Canal near Crofton

There is also a wealth of nature to be seen, from the strange sarsen-strewn landscape of the Fyfield Down National Nature Reserve, to numerous smaller, local reserves cared for by the three wildlife trusts that operate within the AONB: the Berkshire, Buckinghamshire and Oxfordshire Wildlife Trust (BBOWT); the Hampshire and Isle of Wight Wildlife Trust; and the Wiltshire Wildlife Trust.

Sites that charge admission include the likes of Basildon House, Highclere Castle and the Crofton Beam Engine and Pumping Station, along with numerous smaller establishments.

Anglers are also well catered for, with prime fly-fishing for brown trout and grayling on the Rivers Kennet, Lambourn and Pang, and coarse fishing on the Kennet and Avon Canal as well as stocked reservoirs and lakes. The Kennet and Avon Canal is also a popular destination for narrow boats, whether travelling the full length of the canal between Reading and Bath, or just enjoying a leisurely day trip from the likes of Hungerford or Pewsey.

Tourism is definitely a growing industry within the North Wessex Downs and is likely to become a more important driver in the local economy over the coming years.

CHAPTER FOUR

Towns and villages

The patchwork landscape of the North Wessex Downs, with its mixture of high open downland, secluded valleys and ancient woodland, is home to a number of picturesque towns and villages with fascinating histories, interesting churches, cosy pubs and 'chocolate box' cottages – let's take a look at some of the highlights.

Main Market Towns

The two main market towns within the North Wessex Downs – Marlborough and Hungerford – are located along the main A4 road that runs east–west through the middle of the region, a road that has been an important trade route for centuries.

MARLBOROUGH

The picturesque market town of Marlborough hides in the Kennet Valley in the north-east of Wiltshire, hemmed in by the rolling contours of the Marlborough Downs to the north and the large expanse of Savernake Forest with its ancient trees to the south-east. The town, once a staging post on the old coaching route from London to Bristol (now the A4), is said to have one of the widest High Streets in Britain, yet there is much more to Marlborough than just a wide street. There are a number of interesting buildings, old coaching inns, an annual jazz festival, a literary festival and a famous college.

▲ The Town Hall sits at one end of Marlborough's famously wide High Street

For the earliest inhabitants we need to look back some 4,000 years or more to the Neolithic period, but to get there we have to take a slightly convoluted route. Marlborough was first recorded in the Domesday Book as '*Merleberge*': the name could be derived from the hill or barrow of Maerla, but who was Maerla and were they really buried there, or did they own the land … the answer is probably lost in the mists of time. One thing we do know is that Marlborough has a mound, located in the private grounds of Marlborough College, and at one time the Normans built a castle on top of it. Yet many have speculated that the mound was much older; myths and legends suggest that it is the final resting place of the Arthurian wizard Merlin – could that explain the town's name as 'Merlin's Barrow'?

Certainly, the town's motto, '*Ubi nunc sapientis ossa Merlini*', or 'Where now are the bones of wise Merlin?', gives some credence to the legend that has convinced many over the years that Merlin's bones were indeed buried under the mound. The Marlborough Mound has several similarities to that most enigmatic of mounds – the Neolithic Silbury Hill; both have a similar shape, although the mound at Marlborough is smaller, being just 65 feet high with a diameter of 330 feet; both are man-made and both are located close to natural springs along the River Kennet, an area which has a great number of prehistoric sites.

Anyway, enough of legends – let's look at the facts. Recently, the Marlborough Mound Trust, in partnership with English Heritage archaeologists, set about drilling two bore holes down through the mound to obtain samples from within the centre that could be carbon-dated. The results all gave a consistent figure suggesting that the mound was built sometime around 2400 BC, finally establishing once and for all that the Marlborough Mound was built by our Neolithic ancestors at a similar time to Silbury Hill. At two-thirds the height of Silbury Hill, the Marlborough Mound has now been confirmed as the second largest prehistoric mound in Britain. Jim Leary, the English Heritage archaeologist who led a recent excavation of nearby Silbury Hill, said: 'This is an astonishing discovery. The Marlborough Mound has been one of the biggest mysteries in the Wessex landscape. For centuries, people have wondered whether it is Silbury's little sister, and now we have an answer.'

The next main event in Marlborough's history was the arrival of the Romans. They built a settlement – *Cunetio* – to the east of Marlborough on the opposite side of the River Kennet from the present-day village of Mildenhall.

Moving forward to the years shortly after the Norman Conquest of 1066, we know that the Normans reused the Marlborough Mound (rather than actually building their own) as the basis for their motte and bailey castle, digging a moat round the mound and surrounding land. The original castle would have been constructed of wood before being replaced by a more substantial stone structure. From this time on, Marlborough's importance grew and by the early twelfth century Henry I held court at '*Maerlebeorge*' and Savernake Forest became a favoured royal hunting ground. The castle reached the peak of its importance during the reign of Henry III, with Parliament enacting 'The Statutes of Marlborough' there in 1267; some of the laws passed then are still in place today.

History was not kind to Marlborough's castle and by the start of the fifteenth century it was in a ruinous state, yet the ruins were still said to be present over a hundred years later. Today there is little or no trace above ground that a castle ever existed – most likely the stone was reused in some of the town's other buildings. The estate stayed in Royal hands until the time of Henry VIII, but when he married Jane Seymour the estate was given to Jane's brother, Edward, who became the 1st Duke of Somerset; the Seymours built a family home close to the site of the old castle.

By the early eighteenth century a new house had been built and this became the home of the Duke of Somerset's son Algernon, Lord Hertford, who later became the 7th Duke of Somerset. The mound was incorporated into a formal garden with a grotto at the base; by 1723, William Stukeley's drawings show the layout of an extensive formal garden based around the mound and, in 1724, Daniel Defoe mentions the mound when writing about the garden: 'In Marlborough, stands the Duke of Somerset's garden, and is, by that means, kept up to its due height. There is a winding way cut out of the mount, that goes several times round it, 'till insensibly it brings you to the top, where there is a seat, and a small pleasant green, from whence you look over great part of the town'.

Following the 7th Duke's death in 1750, the house was leased as a fashionable coaching inn, known as the Castle Inn, for travellers following the forerunner of the A4; over the next ninety years famous guests included the Prime Minister, William Pitt the Elder (1st Earl of Chatham), in 1767, and the Duke of Wellington. However, the advent of the railways in the 1840s brought about a rapid decline in the coaching trade and the inn closed in 1843.

In the same year a group of clergymen from the Church of England, with the backing of the Archbishop of Canterbury, were looking to found a boarding school whose main purpose would be educating the sons of clergy. On learning that the Castle Inn at Marlborough had become vacant, they took a lease on it and opened Marlborough College in August 1843 with the admission of its first 199 boys. Within five years the school had grown to accommodate 500 pupils; however, conditions were very basic, which led to the 'Marlborough Rebellion' in 1851. This in turn led to a decline in pupil numbers, a rise in debt, and the resignation of the first Master of the College, Matthew Wilkinson. Fortunately for the college the next two Masters (George Cotton from 1852 to 1858 and George Bradley from 1858 to 1870) brought about much-needed reform, allowing the college's reputation to flourish.

More recently (in 1968) the college became one of the first traditional boys' boarding schools to admit girls into the sixth form and, in 1989, the college went fully co-educational with the admission of girls into the Lower School. Today, the college caters for nearly 900 pupils, of which just over a third are girls.

Over the years, Marlborough College has had many famous pupils, including Kate Middleton – now HRH The Duchess of Cambridge, wife of HRH The Duke of Cambridge, the artist and writer William Morris, the poets Siegfried Sassoon and Sir John Betjeman, the yachtsman Sir Francis Chichester (who circumnavigated the globe in *Gypsy Moth VI*), the musician Chris de Burgh, Anthony Blunt (art historian and communist spy – whose ashes, incidentally, were scattered on Martinsell Hill overlooking the Vale of Pewsey), and the engineer Sir Nigel Gresley – designer of famous steam trains including the *Flying Scotsman* and *Mallard*, to name but a few.

In 1204 the town was granted a Royal Charter by King John allowing it to hold a market, and this takes place twice-weekly in the High Street, which is surrounded by Georgian and Tudor buildings and has a church at each end. At number 132 High Street is the Merchant's House, a fine example of a brick and timber Cromwellian town house with tile-hung façade, overlooking the famously wide street. Within the house, which was built in the second half of the seventeenth century by Thomas Bayly, a silk merchant, are several interesting features: the panelled chamber with oak-panelled walls, unchanged since the days of Oliver Cromwell, the dining room painted with a striking

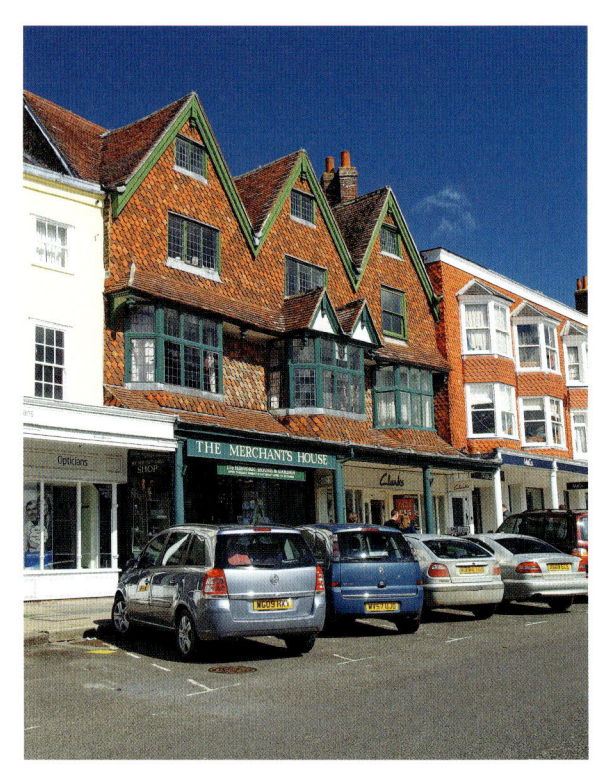

▲ The tile-hung façade of the Merchant's House in Marlborough

vertical strip design from 1665, and the great staircase which dates from the period of Charles II – only recently was a reflected painted image of the balustrade discovered on the stairwell walls.

Owing to its former role as a staging post on the old coaching route from London to Bristol, the town has a number of former coaching inns. These include the Castle & Ball with its tile-hung façade looking out over the High Street; the inn, which was rebuilt around 1745, originally dates from the fifteenth century and was once known as the Antelope. The Sun Inn dates back to a similar time, while the Lamb Inn just off the High Street dates back to 1672.

At the eastern end of the High Street, beyond the Victorian Town Hall, is St Mary's Church with The Green, the site of the original Saxon village further east. There was probably a Saxon church on the site of the present building; certainly there was a church in the late eleventh century and this was replaced with a Norman structure in the mid-twelfth century, though only a few parts of this building remain today. In the fifteenth and early sixteenth centuries the aisles were rebuilt and extended and a crenellated west tower added.

This church was one of many buildings affected by the Great Fire of Marlborough which, in 1653, burnt around 250 houses to the ground. Rebuilding was ordered by Oliver Cromwell, who levied a financial contribution from every parish in the land to help with the cost in gratitude for the town's support for his cause during the Civil War. The church itself was partly rebuilt and the building was modified again during Victorian times. During the seventeenth-century rebuilding a defaced statue of the goddess *Fortuna* was built into the west wall of the nave – maybe the statue came from the nearby former Roman town of *Cunetio*. If you take a look at the north side of the church tower you can still see the marks of shots from Royalist guns; during the Civil War, when Marlborough was under a Royalist siege in 1642, the Parliamentarian commander had taken refuge in the church.

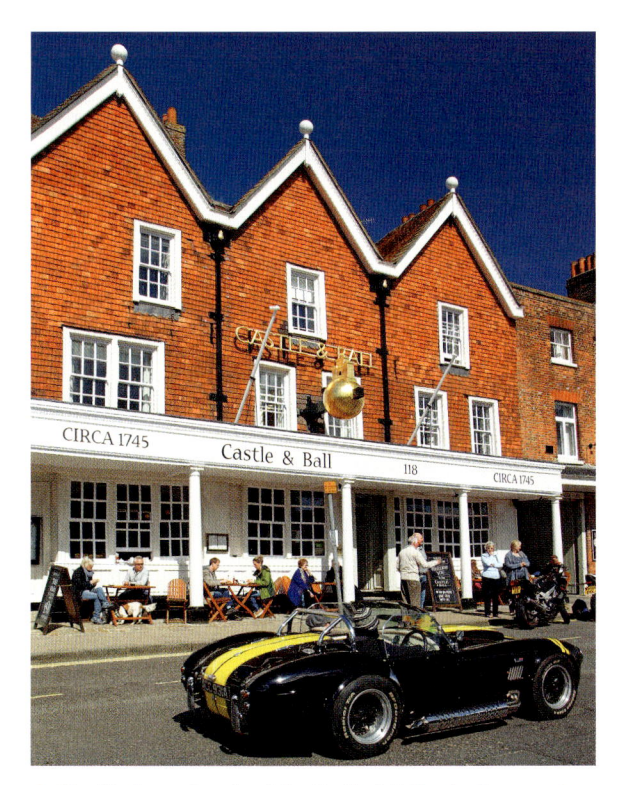

▲ The tile-hung façade of the Castle & Ball, rebuilt around 1745, looks out on the High Street in Marlborough

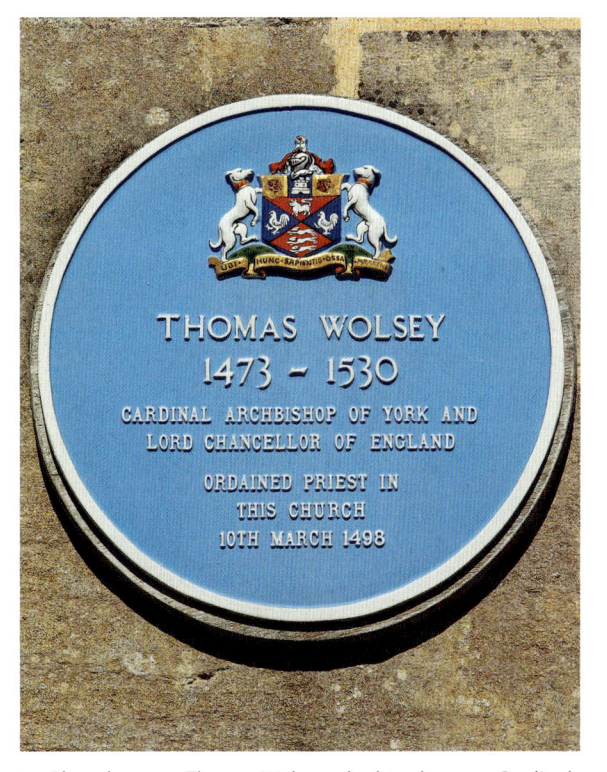

▲ Blue plaque to Thomas Wolsey who later became Cardinal Wolsey; the plaque is one of several in the town

At the western end of the High Street is the Church of St Peter and St Paul. The present building dates from 1460 and stands on the site of an earlier church, which had been built to serve the nearby castle. In 1498 Thomas Wolsey (1473–1530) was ordained as a priest in the church – an event commemorated by a blue plaque. He later rose to become Cardinal Wolsey and Lord Chancellor for Henry VIII, before failing to get an annulment for Henry VIII's marriage to Catherine of Aragon so that Henry could marry his second wife Anne Boleyn; this state of affairs led to England splitting from Rome.

The church, as with many churches, underwent a major Victorian restoration in 1863 and finally closed as a parish church in 1974. The building remained empty until the St Peter's Trust was formed in 1978 to safeguard the building from demolition; the church is now run as a community centre and café.

Throughout the town there are several other blue plaques commemorating people who have lived in Marlborough.

The author and Nobel Literature Laureate William Golding (1911–1993) grew up at 29 The Green while his father was a science master at Marlborough Grammar School. Golding, best known for his novel *Lord of the Flies* (1954), renamed Marlborough as *Stilbourne* and made it the setting for his 1967 novel *The Pyramid*.

At 3 and 4 High Street is a plaque to two inventor brothers, Thomas and Walter Hancock. Thomas Hancock (1786–1865) patented the process known as vulcanization in November 1843, whereby sticky natural rubber is converted into a more durable and usable product with the addition of sulphur. He later went into partnership with a Scot called Mackintosh making 'macs' that are still worn today. His younger brother Walter Hancock (1799–1852) was noted for inventing the steam passenger road carriage.

Further along the High Street, on the front of the library, is a plaque recording that Eglantyne Mary Jebb (1876–1928), founder of the Save the Children Fund, taught there when it used to be St Peter's School, while at 48 High Street a plaque marks the former home of a tanner where it's believed the Great Fire of 1653 started.

HUNGERFORD

Moving east from Marlborough along the A4 brings us to Hungerford, whose name is most likely derived from a Saxon name meaning 'hanging wood ford'. The fourteenth-century chronicler of the *Book of Hyde* mentions that a Dane – Hingwar, often referred to as Ivarr the Boneless – drowned crossing a river at a place called 'Hingwar's Ford', while on his way to meet the Saxons in battle at Ethandune. Some have claimed that the battle took place at nearby Eddington; however, modern opinion suggests that it actually took place at Edington in Wiltshire, so it is unlikely that Hungerford's name has anything to do with Hingwar.

Little is known about the town's early history, although Neolithic stone tools and a Bronze Age building were found close by in 1989, and a former Roman road passed just to the north. The Domesday Book mentions many of the surrounding manors but not Hungerford itself. The first written evidence dates from 1108, when documents refer to a church of Hungerford being assigned to the Abbey of Bec-Hellouin in Normandy.

The manor of Hungerford passed between the Crown and various duchies of Leicester and Lancashire between the eleventh and fourteenth centuries and in 1362 the manor passed to the wife of Prince John of Gaunt, who was the fourth son of Edward III. It was John of Gaunt who granted the rights of free grazing and fishing to local 'commoners'.

In 1612, James I granted the Manor of Hungerford to two local men, who in turn passed on the responsibility to a group of fourteen *feoffees*, or trustees, in 1617. The present commoners are those people owning and living in the properties established at the time of the 1612 grant.

The organization of the town has remained virtually unchanged for 400 years with office holders including the Constable, Water-Bailiffs, Overseers of the Common (the common being Port Down), Keepers of the Keys of the Common Coffer, Ale-Tasters and the Bellman – although some of the old officers' titles, such as the Searchers and Sealers of Leather and the Tasters of Flesh and Fish, have disappeared over the years.

▲ The site of the former wharf on the Kennet and Avon Canal at Hungerford

An important event in British history took place in Hungerford on 8 December 1688. A group of leading political figures of the day, having become disenchanted with the rule of the Catholic King James II, sent an open invitation to the Protestant Prince William of Orange. He set sail for Brixham in Devon and started to march towards London. On reaching Hungerford, where he stayed at the Bear (now the Bear Hotel), he was met by three commissioners sent by James II. The soldiers of James II had already started to switch their allegiance to William, and James II fled to France, opening the way for William to jointly rule as William III with his wife Mary II (daughter of Charles II); a plaque on the wall of the hotel commemorates the historic meeting.

Some years earlier, during the Civil War, the Bear Hotel had received another royal visitor when Charles I stayed there following his retreat from the Second Battle of Newbury in 1644. It is, in fact, one of the oldest and most historic inns in England, having been first mentioned in 1494, although it is reputed to date back to the thirteenth century. The hotel's name and sign are derived from the 'Bear and Ragged Staff' badge of the Earls of Warwick, owners of the Manor of Chilton Foliat in the fifteenth century, to which the Bear belonged. The Coat of Arms over the entrance on Charnham Street is that of the Popham family, Lords of the Manor of Chilton Foliat from 1607 until 1929.

Hungerford has, for many centuries, had good travel links. The old coaching road – the Bath Road – passed just north of the town on its way from London to Bath and Bristol, while the north–south route between Oxford and Salisbury also passed by. The horse-drawn coaches travelling along these routes brought an increasing level of prosperity to the town, whether through the large number of coaching inns, the stables, or the blacksmiths who serviced the travellers and their horses. In 1810 the Kennet and Avon Canal arrived in town, crossing under the High Street at its lowest point, while just to the south is the bridge for the railway between Reading and Westbury. The line from Reading to Hungerford, which was built by the Great Western Railway-backed Berks & Hants Railway, opened in 1847.

Further up the High Street, which is home to a weekly market, is the Victorian Town Hall built in a

▲ The Byzantine-styled Town Hall in Hungerford

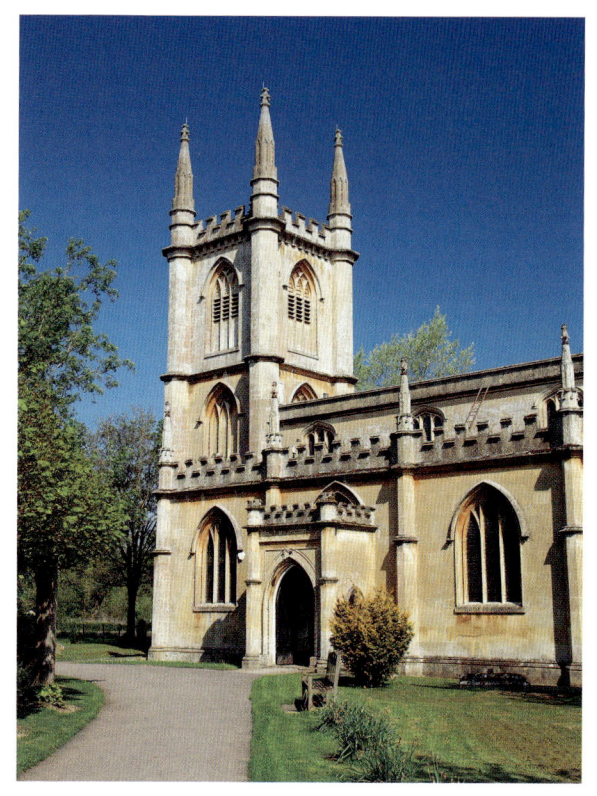

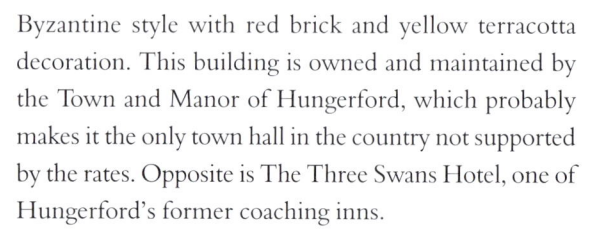

▲ The Church of St Lawrence, made from Bath stone transported along the Kennet and Avon Canal

Byzantine style with red brick and yellow terracotta decoration. This building is owned and maintained by the Town and Manor of Hungerford, which probably makes it the only town hall in the country not supported by the rates. Opposite is The Three Swans Hotel, one of Hungerford's former coaching inns.

Slightly away from the town centre, close to the Kennet and Avon Canal and adjacent to the River Dun, is the Church of St Lawrence, which was built in 1816 to replace an earlier church. Designed in a Gothic Revival style by James Pinch of Bath, the church was constructed from Bath stone that was brought to the site using the newly opened Kennet and Avon Canal. Inside is the rather mutilated effigy of Sir Robert de Hungerford (d.1352), who built a chantry chapel in memory of his wife; the family, named after the town, were living there as early as 1165.

Between this church and the High Street is the Church Croft or village green, around which the original village grew. The land was given to the people of Hungerford 'to hold fairs and sport therein' by John Undewes and his wife in the mid-sixteenth century at the nominal rent of a single red rose yearly. The land came under the control of the town trustees in the Feoffment of 1617.

To the east of the town is the common or, to give it its full title, Hungerford Common Port Down. Fortunately common rights have been preserved over the centuries, with cattle still grazing on the common, and this has helped prevent the spread of invasive scrub and tree growth that would otherwise have altered its character.

To the west of the town, bisected by the River Dun and Kennet and Avon Canal, is another area of common land, known as Freeman's Marsh and designated a Site of Special Scientific Interest (SSSI). The land to the south of the canal is grazing land, while to the north is open water meadow through which the River Dun flows, making it a great place for wildlife.

A special day in Hungerford's calendar occurs each year on the second Tuesday after Easter, when the town celebrates Tutti Day which forms part of the Hocktide Festival marking the end of the financial year; Hungerford is the only place in England to still hold this annual festival. The day starts with the Bellman summoning the Commoners, of which there are close to a hundred, to the Hocktide Court. Here, the accounts are read and agreed, the officers for the coming year are elected, and any other matters concerning the affairs of the Town and Manor are also debated. The Town and Manor became a registered charity in 1908 and the trustees cover the running of the Town Hall, the early seventeenth-century John of Gaunt Inn, Common Port Down, Freeman's Marsh and the fishing rights along parts of the River Kennet and Dun.

While the court is in session, two Tutti-men carrying tutti-poles – wooden poles topped with flowers and a cloved orange – work their way around the town, visiting every house with common rights. In days gone by, this was when the rent or 'head penny' would have been collected from each householder. These days what they usually collect is a kiss from the ladies of the house and some hospitality to help them on their way.

Kennet Valley Villages

RAMSBURY

Lying on the northern bank of the River Kennet between Marlborough and Hungerford, not far from Littlecote House and its Roman Orpheus Mosaic, is Ramsbury, whose Anglo-Saxon name is derived from 'the fortified place of the raven'. At one time the village was on the main coaching route between London and Bath. However, the present A4 takes a more direct route rather than following the river valley. During the Second World War there was an RAF airfield to the south across the River Kennet which became operational in 1942 (see North Wessex at War – page 56).

According to John Aubrey, who passed through one April in the seventeenth century, the river was used to create water meadows that were 'yellow with butter flowers'. Later in the nineteenth century the clear waters of the Kennet were used to grow watercress. Ramsbury also had a tradition for making beer; back in the late eighteenth century it was mentioned in the Wiltshire Directory that 'Ramsbury is noted for the excellent beer of which there is a great consumption in London', a tradition that has been recently revived with the opening of Ramsbury Brewery.

The village can trace its history back to the Saxons, when the Bishopric of Ramsbury was created in AD 909 (the 'See' was later moved to Old Sarum in 1075); the remains of a Saxon iron foundry have also been found where the High Street now lies. The estate remained for centuries in the possession of the Bishops of Salisbury, but in 1545 it was granted by Bishop Salcot to Edward Seymour, Earl of Hertford, passing to the Duke of Somerset in 1547 and then to William Herbert, Earl of Pembroke, in 1552. It was later sold to Henry Powle in 1676–7 and in 1681 Powle sold it to Sir William Jones (d.1682), who was the Attorney General under Charles II. He had the manor house (situated to the west of the village) rebuilt by Robert Hooke in the English Renaissance style on the site of an earlier house that had been built by the Earl of Pembroke in 1560. The house then passed through several members of the Jones family before passing to Sir Francis Burdett (d.1844), a Whig politician, on the death of Lady Jones in 1800; the manor stayed in the Burdett family for the next 150 years.

In the centre of the village is the Church of the Holy Cross. The current building dates from the thirteenth century, yet there was a 'large and elaborate' church on this site as far back as the start of the tenth century, during the time of the Bishopric of Ramsbury. The dedication of the church probably dates from the mid-thirteenth century when two fairs were granted at the Invention and Exaltation of the Cross (3 May and 14 September) to replace the original market. The sturdy tower with corner buttresses was built in the fourteenth century and the clerestory was added in the early sixteenth century; extensive renovations were undertaken in the late nineteenth century.

Inside the church is an interesting collection of memorials including one to Sir William Jones, former owner of Ramsbury Manor. There are also memorials to

▲ Part of a Saxon carved cross in the Church of the Holy Cross, Ramsbury

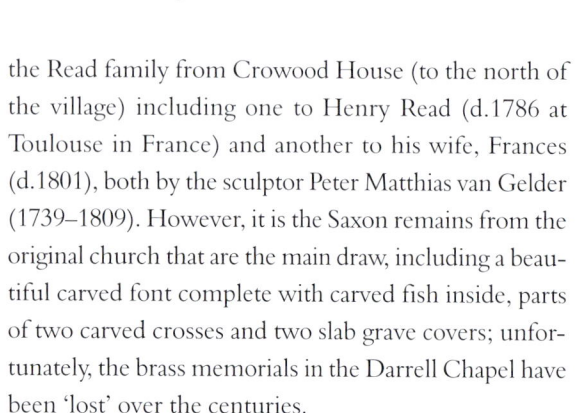

▲ Part of a Saxon carved grave cover in the Church of the Holy Cross, Ramsbury

the Read family from Crowood House (to the north of the village) including one to Henry Read (d.1786 at Toulouse in France) and another to his wife, Frances (d.1801), both by the sculptor Peter Matthias van Gelder (1739–1809). However, it is the Saxon remains from the original church that are the main draw, including a beautiful carved font complete with carved fish inside, parts of two carved crosses and two slab grave covers; unfortunately, the brass memorials in the Darrell Chapel have been 'lost' over the centuries.

ALDBOURNE

To the north, tucked away from the River Kennet, is Aldbourne; the name is derived from the Saxon word for stream – bourne – and a former local chief – Ealda. The interesting Church of St Michael, on the village green, was rebuilt in a Gothic style in the thirteenth century, incorporating some Norman features from an earlier church that had been built in the twelfth century; the Perpendicular-styled tower was added around 1460.

▲ The Saxon font in the Church of the Holy Cross, Ramsbury

▲ The village green and the Church of St Michael in Aldbourne

Inside there are some fine memorials, including one to Richard Goddard (d.1492) and his wife Elizabeth of Upham House. There are also two old fire pumps, known locally as Adam and Eve. Originally bought after a disastrous fire in 1777, which damaged many houses in the village, the pumps were last used for a barn fire in 1921.

The village was once well known for its bell foundries, with the first foundry being opened in the late seventeenth century by the Cor family; their trademark was the emblem of a bird. In the eighteenth century Robert Wells opened another foundry which continued operating into the nineteenth century. It was once said that there were few places in north Wiltshire that were out of earshot of a bell cast in Aldbourne.

Of interest to all *Dr Who* fans is the fact that the village was transformed into *Devil's End* for the 1971 BBC TV series *The Daemons*. The Blue Boar pub was renamed *The Cloven Hoof* and during the final episode the church was blown up – albeit this was just a realistic model.

Interestingly, locals were once known as 'Aldbourne dabchicks'. The story goes that a dabchick, or little grebe, was once seen on the village pond but the locals didn't know what it was; this amused the people of Ramsbury, who taunted their Aldbourne neighbours by calling them 'dabchicks'. Nowadays, the name 'dabchick' is used by several groups within the village including the rugby club. It is also the name of the village magazine.

KINTBURY

Moving eastwards along the Kennet Valley brings us to Kintbury. The first written evidence of the village, whose name means 'fort on the River Kennet', was in a Saxon document in which it was known as *Cynetanbrig*. At that time Kintbury was the centre of a 'Hundred', a Saxon administrative area predating parishes – though by the time of the Domesday Book it was known as *Chenetbrie*. However, the area's history goes back much further: evidence of a Roman settlement was unearthed at Irish Hill to the east of the

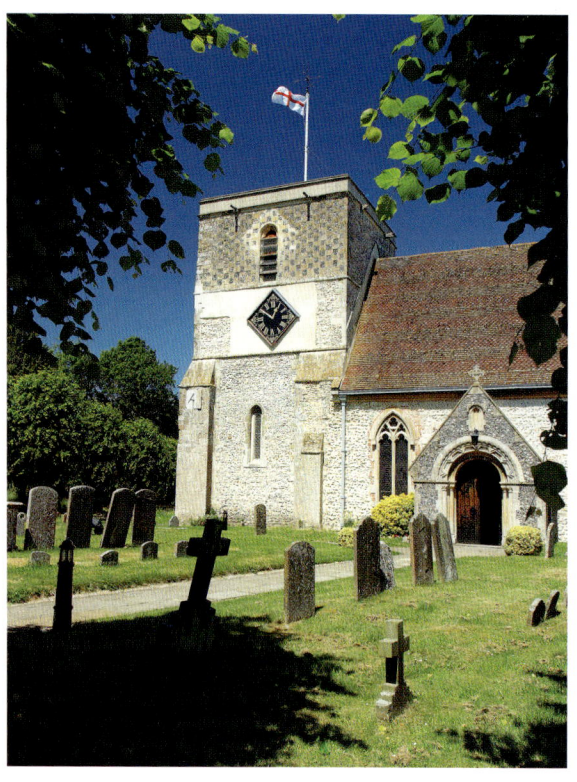

▲ St Mary's Church, Kintbury

village and 5,000-year-old worked flints have also been found.

The chancel and nave of St Mary's Church date from the early twelfth century and there are two good examples of Norman doorways. The attractive chequerwork extension to the tower was added in the fifteenth century. Inside the church are a number of interesting monuments, including a brass memorial in the chancel to John Gunter (d.1624) and his wife Alice (d.1626), while on the west wall of the tower there is a royal coat of arms of Charles II from 1683. In the north transept are monuments to three generations of the Raymond family who inherited Barton Court: one of these bears the name of the famous Flemish sculptor Scheemaker.

Barton Court sits to the north of the River Kennet and was the main house for the manor of Kintbury-Amesbury, so named because Amesbury Abbey owned the lands. The present house was built in the late seventeenth century.

A local legend mentions the 'Kintbury Great Bell' which supposedly sank into the Kennet following a great storm that destroyed the church tower, causing the bell to roll down to the river; another version claims that the bell sank into the river while being transported to the church. Many attempts were made to retrieve the bell, but all failed. A local wizard was consulted and stated that the bell could only be recovered by twelve white heifers driven by twelve maidens in white, in total silence at midnight. At the appointed hour the heifers began to haul on the chain and the bell started to rise from the river, but someone – maybe it was the 'Kintbury Witch' – cried out, the chain broke and the bell sank back into the river where it has remained hidden ever since.

Take a look round the village: many of the bricks used in the old cottages were probably made in the large brickworks that were sited to the south of the village. Down from the village is the Kennet and Avon Canal that opened in 1810, being followed by the GWR-backed Berks & Hants Railway from Reading to Hungerford in 1847.

BEENHAM

Continuing east along the Kennet Valley from Kintbury brings us to Beenham. The earliest written mention of the village was in Saxon times when, in AD 956, it was referred to as *Benna's Hamme*, meaning Benna's meadow or enclosure. Later references were made to a manor of Beenham being granted by Henry I to Reading Abbey on its foundation in 1121. However, following the Dissolution of the Monasteries, Henry VIII granted the manor of Beenham to Sir Henry Norreys.

The parish church of St Mary's is built on the site of two previous churches. The first, of Saxon origin, was destroyed by fire in 1794 after being struck by lightning. Unfortunately the second church was also destroyed by fire (except for the brick tower) and replaced by the present building in 1859. The religious murals in the nave were painted by Mary Sharp of Ufton Court with the help of a London painter in the late nineteenth century. She also painted *The Last Supper* (1879) in the sanctuary (the part of the church containing the altar).

TOWNS AND VILLAGES

▲ One of the late nineteenth-century murals painted by Mary Sharp inside St Mary's Church, Beenham

Just to the north of the village is Butlers Farm, home of the Wolf Conservation Trust, where wolves from various parts of the world can be seen (see also page 166).

Pang and Thames Valley Villages

BUCKLEBURY

The picturesque village of Bucklebury, originally recorded as *Borchedeberie* in the Domesday Book, is situated next to the River Pang. The village rose in prominence when Reading Abbey, who held the manor, built a manor house here for the abbot. However, with Henry VIII's Dissolution of the Monasteries, the manor was sold to one John Winchcombe, son of the famous Jack of Newbury (also known as John Winchcombe, alias Smallwood), who had made his wealth in the cloth trade. By 1703 the male line of the family had died out, so the manor passed to Frances Winchcombe, wife of Henry St John, 1st Viscount Bolingbroke, who later

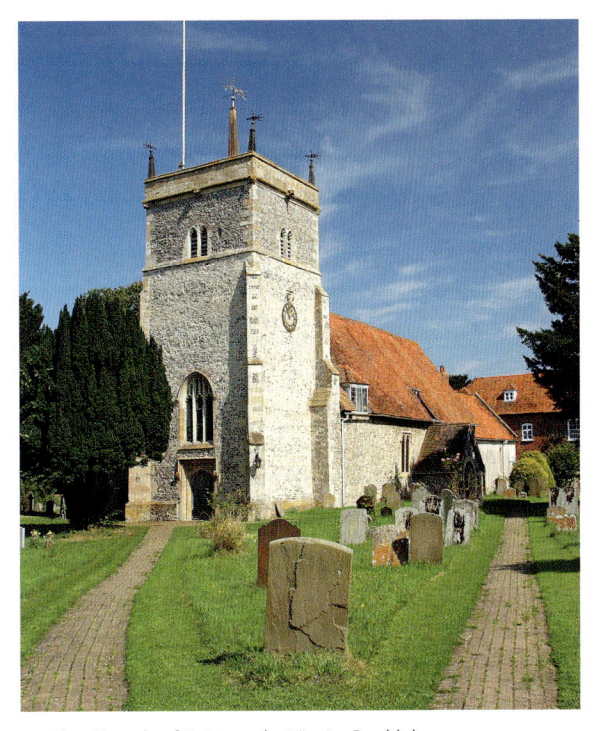

▲ The Church of St Mary the Virgin, Bucklebury

93

▲ This beautifully carved Norman doorway leads into the Church of St Mary the Virgin in Bucklebury

deserted his wife and fled to France. Unfortunately the Elizabethan manor house was extensively damaged by fire in 1830 and was later demolished, apart from one wing, which forms part of the current Bucklebury House.

The eleventh-century Church of St Mary the Virgin has some impressive features, including an elaborately carved Norman south doorway. Interior features include the high Georgian box pews, and six interesting hatchments – coats of arms on diamond-shaped frames. The east window in the chancel depicting the Crucifixion, by Sir Frank Brangwyn (1867–1956), is unusual for both its use of striking colours and in that the crucified Christ is looking up to heaven, rather than down at the ground. While looking round, make sure you see the curious 'fly-window' above a recessed pew in the chancel. The small stained-glass window, dated 1649, has a square panel painted with a sundial. Unfortunately the dial has lost its gnomon (the bit that casts the shadow), yet even if it were still there the sundial would not work in its present location. The realistic fly is believed to be a pictorial substitute for the usual sundial motto – *tempus fugit*, or 'time flies'.

More recently, the area around Bucklebury became known as 'Kate Middleton Country' with the wedding of HRH Prince William, Duke of Cambridge, and Catherine Middleton on 29 April 2011 at Westminster Abbey in London; Catherine's family home is in Chapel Row, which lies within the parish of Bucklebury.

BRADFIELD

Heading downstream along the River Pang from Bucklebury is Bradfield, whose name simply means 'broad field'. The village was first recorded in a late seventh-century Saxon charter which mentions that King Ini, 'the lawgiver of Wessex', granted land at *Bradanfelda* to the Abbey at Abingdon; by the time of the Domesday Book the land was in the ownership of William FitzAnsculf.

Much of what you can see in the village today is a result of the efforts of one man; the Revd Thomas Stevens, a former Victorian rector. St Andrews Church, which dates from the early 1300s, was originally dedicated to St John the Baptist, being rededicated to St Andrew in 1848 after extensive enlargement undertaken by Stevens, with the help of his friend, the leading Victorian Gothic Revival architect Sir George Gilbert Scott (1811–1878). Scott was responsible for designing such famous structures as the Midland Grand Hotel at St Pancras Station and the Albert Memorial, both in London. However, having enlarged the church so much, Stevens found it difficult to fill all the pews, so he decided to establish St Andrews School (now Bradfield College) in 1850 to provide a choir and larger congregation. As the school grew to dominate the village, and Stevens' other ventures drained his financial resources, he was eventually declared bankrupt and the school was taken over by a board of trustees, who employed Dr Herbert Gray as headmaster. It was Dr Gray who turned Bradfield College into one of the leading schools in the area.

Famous pupils have included Richard Adams (b.1920) – author of *Watership Down*; Henry Pelham Lee (1877–1953) – internal combustion engine pioneer and founder of the Coventry Climax Engines company; David Owen, Baron Owen (b.1938) – former Foreign

Secretary and co-founder of the Social Democratic Party (SDP); and Stephen Coleridge (1854–1936) – author and co-founder of the National Society for the Prevention of Cruelty to Children.

TIDMARSH

After passing through Bradfield, the River Pang heads north towards Pangbourne, following the A340 through Tidmarsh. This road was a turnpike road with its own toll house until 1872; the hexagonal brick building, now a private house, stands close to the road in the centre of the village. During the Second World War the valley heading north to the River Thames at Pangbourne formed part of the GHQ Stop Line Red (see North Wessex at War – page 53), which continued upstream towards Abingdon.

The twelfth-century Church of St Laurence has a mixture of Norman and Early English styles; the carved doorway is a particularly fine example of Norman architecture and inside there is an arch-decorated Norman font. Other notable features include the ceiling of the chancel made from panelled oak and the vivid stained-glass window in memory of the physician Harold Kingston Graham-Hodgson (1890–1960). In the floor at the east end of the nave are three tomb slabs with ancient brasses: one to Margaret (d.1499), wife of Thomas Wode, a justice of Common Pleas, and formerly the wife of Robert Lenham, Lord of the Manor of Tidmarsh; one to William Dale (d.1533) and his wife Elizabeth; and a third depicting a knight in sixteenth-century armour. During Victorian restorations many early wall paintings, probably from the mid-thirteenth century, were discovered. However, being imperfect, most were plastered over, with the exception of those on the jambs of the north-east window of the nave.

▲ Vivid stained glass memorial window in the Church of St Laurence (Tidmarsh) to the physician Harold Kingston Graham-Hodgson

To the north of the church is The Greyhound pub which occupies a medieval timber-frame cruck house and has been an inn since 1625. Close by on the River Pang there was a corn mill for over 700 years, first mentioned in 1239 and held by the Abbot of Reading until the sixteenth-century Dissolution of the Monasteries. The present Mill House, most likely located on the site of the original mill, was once the home of Dora Carrington and Lytton Strachey, both members of the Bloomsbury Set (see page 145).

SULHAM

Just to the east of Tidmarsh, across the River Pang, is Sulham. Before the Norman Conquest the manor was held by Godric, although by the time of the Domesday Book the manor had passed to William de Calgi. For the next few hundred years the manor passed through different ownership until 1632 when it was bought by the Wilder family from nearby Nunhide; the family still own the manor.

The Norman church was replaced with the present small, but impressive, parish church of St Nicholas in the 1830s by the Revd John Wilder (1801–1892), one of many Wilders who became rectors of the parish in an unbroken line from 1823 until 1944. The adjacent Sulham House was built in 1701 during the reign of William and Mary, although it has undergone many alterations, the last being in the 1830s when the church was rebuilt.

Wilder's Folly (also known as the Nunhide or Pigeon Tower) to the south of the village was built by the Revd Dr Henry Wilder (see page 140).

PANGBOURNE

The popular riverside town of Pangbourne dates back at least to Saxon times, having been mentioned in a Saxon charter of AD 844, when Ceolred, Bishop of Leicester, granted land to Berthwulf, King of Mercia. At this time the village was known as *Paegingaburnum*, meaning the 'stream of Paega's people'. At the junction of St James' Close and Station Road (next to the car park) is the village sign which depicts two local heroes. The first is Berthwulf, King of Mercia, along with the village charter and a Saxon ship above the name of the village. The other, separated by more than 1,000 years, is Kenneth Grahame (1859–1932) – symbolized by an open book and trees. Grahame, who lived in Church Cottage next to the church from 1924 until 1932, wrote the children's classic, *The Wind in the Willows*.

Pangbourne has long been a favoured crossing point of the River Thames. Originally a ferry operated from Ferry Lane in Pangbourne to the neighbouring Oxfordshire village of Whitchurch. However, following the Whitchurch Bridge Act 1792, the ferry was replaced with a wooden bridge; this was subsequently replaced with a second wooden bridge in 1853 and that, in turn, was replaced by the present iron bridge in 1902. Tolls are still collected from those driving across the bridge to pay for its upkeep. In 1840 the arrival of Isambard Kingdom Brunel's famous Great Western Railway en route from London via Reading and Swindon to Bristol brought with it new prosperity.

Throughout its history the town has had many famous inhabitants and visitors. Sir Benjamin Baker (1840–1907), designer of the Forth Rail Bridge (1890) and the first Aswan Dam (1902), lived in the village. It was also while at the Swan Inn, which dates from the 1640s, that Jerome K. Jerome's *Three Men in a Boat* finally abandoned their adventure along the River Thames.

The Parish Church of St James the Less, one of fewer than thirty churches in England dedicated to that saint, was built on the site of an earlier church some time prior to 1175. The huge red brick tower was added in 1718, while the nave and chancel were rebuilt in the 1860s. Inside there is a lovely modern window in the north wall depicting St James the Less and St Cecilia, along with seven hatchments from the Breedon family, long-standing lords of the manor at Bere Court. Probably the building's greatest treasure, though, is the huge east window by Karl Parsons. This beautiful window, rich in colour and detail, depicts the Virgin and Child along with St George and St Michael, and is claimed to be one of the best examples of his work. The window was given by Sir George and Lady Armstrong in memory of their son and all men of the parish who died in the Great War (1914–18). A portion of the window, depicting the Angel Gabriel, was used on the Royal Mail Second Class Christmas stamp in 1992.

Pangbourne Manor, which included Bere Court to the south-west of the village, was given to the Abbey of Reading by Henry I, remaining in that ownership until Henry VIII's Dissolution of the Monasteries, when the manor passed to the Crown. In 1613 Bere Court was bought by Sir John Davis (1560–1625) who fought in

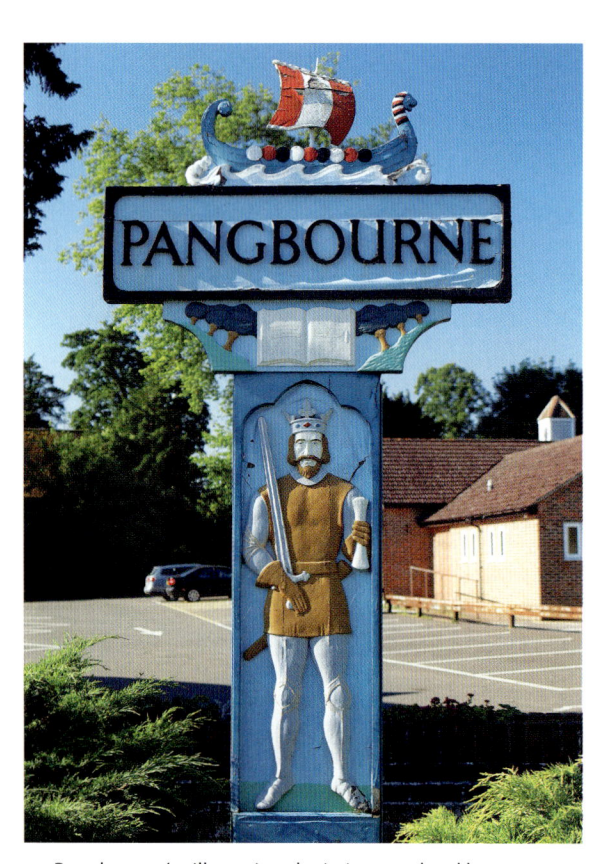

▲ Pangbourne's village sign depicting two local heroes

▲ The Bull at Streatley where Jerome K. Jerome's Three Men in a Boat had lunch

the Spanish Wars under Robert Devereux, 2nd Earl of Essex; Davis was knighted after the taking of Cadiz in 1596 and inside St James' Church there is a fine effigial monument to Sir John.

Bere Court was purchased by the Breedon family in 1671 and they continued to be lords of the manor until the late nineteenth century when it was sold to Reginald de la Bere. The present Bere Court, an early Georgian brick house, was built for John Breedon, one-time Sheriff of Berkshire (d.1711).

Pangbourne College, or The Nautical College as it was originally called, was founded by Sir Thomas Lane Devitt in 1917 with the purpose of preparing boys to be officers in the Royal Navy and Merchant Navy; the name Pangbourne College was adopted in 1969. Within the college grounds is the Falkland Islands Memorial Chapel opened by Queen Elizabeth II in March 2000. The chapel was built to commemorate the lives and sacrifice of all those who died during the Falklands War of 1982, and the courage of the servicemen and servicewomen who served with them to protect the sovereignty of the Falkland Islands; the chapel is open daily.

STREATLEY

Moving upstream along the River Thames from Pangbourne is Streatley, sitting on the Berkshire side of the Thames with its neighbour, Goring, on the Oxfordshire side. The village has a history dating from Anglo-Saxon times and probably a lot earlier than that, though we do know that, following the Norman Conquest, Geoffrey de Mandeville was lord of the manor. Streatley House in the High Street, built in 1765, was once the home of the Morrell family, former brewers in Oxford, while the Bull at Streatley pub was where Jerome K. Jerome's *Three Men in a Boat* had lunch. The Parish Church of St Mary – which dates from the thirteenth century, though it was largely rebuilt in 1864 – houses some interesting fifteenth- to seventeenth-century brasses.

Villages in the Berkshire Downs

WEST ILSLEY

Picturesque West Ilsley with its village pub, The Harrow, overlooking the cricket ground and duck pond, has a gazebo erected by the villagers to mark the new millennium. Further along Main Street are the old village school, rectory and church. All Saints Church underwent major alterations in the 1870s, though the underlying fabric of the church is much older. Inside there is a fine carved wood Jacobean pulpit along with a rood screen and cross. The village was the original home of the Morland Brewery. John Morland, a local farmer, started brewing his much-sought-after ale here in 1711, though the business eventually moved to Abingdon in 1887.

EAST ILSLEY

To the east of the A34 is East Ilsley, once famed for its sheep market which was second only to Smithfield Market in London; at its peak in the 1880s around 20,000 sheep were changing hands in a single day. The last sheep fair was held in 1934, although the village recently started holding an annual Sheep Fair Fête. The village, which nestles round the duck pond, has several interesting buildings including East Ilsley Hall and Kennet House, two impressive early eighteenth-century houses.

St Mary's Church was built sometime in the twelfth century – it is known that in 1199 King John gave the church to the Knights Hospitaller of St John of Jerusalem. However, most of the present building dates from the thirteenth century, with the tower being added in the fourteenth century and restored in 1625. The church is claimed to have one of the finest peals of eight bells in Berkshire; the bells include one by Joseph Carter of Reading (c.1589), one by William Yare of Reading (c.1612), and one by the famous Cor family from Aldbourne (see page 91). A new clock was placed in the church tower in the late nineteenth century. The original clock, reputedly made by a local blacksmith in 1627 and now housed in the Museum of History and Science in Oxford, had no face and only rang on the hour. The clock, plus the large number of public houses that the

▲ The Church of St Mary and St Nicholas in Compton

village used to have – claimed to be thirteen at one time – gave rise to a local rhyme:

Sleepy Ilsley, drunken people,
Got a church without a steeple,
And what is more, to their disgrace,
They've got a clock without a face!

COMPTON

Nearby Compton gets its name from '*cwm tun*' meaning 'valley settlement' – a good description, as the village is 'hidden' in the downs near to where the River Pang rises. The village originally consisted of at least two manors: East and West Compton.

West Compton now forms the bulk of the present village to the west of the disused railway line, which at one time gave Compton a rail link to Didcot and Southampton and was important for transporting sheep to and from East Ilsley; the line closed in 1962.

East Compton, probably the older settlement, was located to the east of the Parish Church of St Mary and St Nicholas and was for several hundred years held by the Abbey of Wherwell in Hampshire until the Dissolution of the Monasteries; much later, in the nineteenth century, Lord Wantage gained both manors through

TOWNS AND VILLAGES

▲ The Pygott brass in the Church of St Mary and St Nicholas, Compton

marriage. Although the church dates from around the thirteenth century, with an Early English-style chancel, it underwent extensive renovations in the nineteenth century. Inside are a circular Norman font and some interesting monuments, including one to Richard Pygott and his wife Alice (*c.*1500).

In 1937 Compton became the home of the Institute of Animal Health when the Agricultural Research Council established a research site there (other sites were based at Houghton and Pirbright); at the time of writing the Compton site is being decommissioned and moved to Pirbright. The institute's farm – Mayfields Farm – is still operating and grows various crops as well as having a herd of Holstein Friesian dairy cows and a flock of Dorset sheep which provide animals of known high health status for research.

The village played a role in the early 1990s BBC series *Trainer*. The series, set in the world of horse racing, was partly filmed in Compton and in neighbouring East Ilsley.

▲ Church of St Peter and St Paul in Yattendon (final resting place of the former poet laureate Robert Seymour Bridges)

YATTENDON

To the south-east of Compton is Yattendon, known as *Etingedene* in the Domesday Book. The manor, held in the fourteenth century by the de la Beche family from nearby Aldworth, later passed to the Norreys family through marriage in the early fifteenth century, and Sir Henry Norreys was granted a licence in 1448 to build and fortify a manor house.

Much later, in 1876, the manor was bought by Alfred Waterhouse, famed for being the architect who designed the Natural History Museum in London. He built his house, Yattendon Court, to the east of the village. It was during the construction of the house that the Yattendon Hoard was discovered in 1878. This consisted of fifty-nine items from the Bronze Age, including arrowheads, knives, parts of swords, axes and various tools. Unfortunately this new house did not survive long as the newspaper proprietor Sir Edward Mauger Iliffe (1877–1960), 1st Lord Iliffe, who bought the estate in 1925, demolished it to make way for the present Yattendon Court. Lord Iliffe formed the Yattendon Estate through the amalgamation of several smaller estates which he acquired between 1925 and 1940, while his son (Edward Langton Iliffe, later the 2nd Lord Iliffe) restored Basildon Park (see Stately Homes – page 118).

The present Church of St Peter and St Paul, built in the Perpendicular style, dates from the mid-fifteenth century, having been rebuilt by the lord of the manor, Sir Henry Norreys, although it is known that there was an earlier church here as the list of rectors goes back to the late thirteenth century. The church, which is the final resting place of the former poet laureate Robert Seymour Bridges (see Writers of Prose and Poetry – page 148), is noted for the Yattendon Hymnal, a collection of ancient hymns edited by the poet Robert Bridges in 1899.

Nearby Hampstead Norreys is home to St Mary's Church and a former Second World War airfield (see North Wessex at War – page 54).

▲ Brick and timber cottage in Blewbury (see opposite page)

Villages in the Northern Part of the AONB

BLEWBURY

Up in the north-east corner of the AONB is Blewbury, once home to Kenneth Grahame (1859–1932), Secretary to the Bank of England – though better-known as author of *The Wind in the Willows*. The village has many picturesque timber-framed and thatched-roofed cottages, and some original thatched cob boundary walls. First mentioned as *Bloebyrig* in a Saxon charter from King Edmund in AD 944, by the time of the Domesday Book it was known as *Blitberie*. By then it was quite large, with a church and four mills.

St Michael's Church, though rebuilt in the eleventh century, has Saxon origins. Inside are some interesting memorial brasses, including one to Dame Alice Daunce (d.1523) and her husband Sir John, Surveyor-General to Henry VIII.

ASTON TIRROLD AND ASTON UPTHORPE

Next door to Blewbury are the twin villages of Aston Tirrold and Aston Upthorpe, which at one time were known for their Presbyterian tradition; 'meetings in the barns' were fairly common following the Act of Uniformity in 1662. In 1717 'there were two hundred hearers' attending the meeting and in 1728 the interesting Presbyterian Chapel, now the United Reformed Church, was built. Aston Upthorpe's All Saints Church is built on Saxon foundations and has an eleventh-century nave and filled Norman doorway, while St Michael's Church, in Aston Tirrold, dates back to 1080. The railway line passing just to the north is Isambard Kingdom Brunel's famous Great Western Railway, built in the 1850s to connect London and Bristol.

Tucked right on the edge of the AONB beside the Cholsey and Wallingford Railway is Cholsey; although the village itself lies outside, the village church – St Mary's – is within the AONB. In the churchyard are the graves of the world-famous crime writer Dame Agatha Christie (1890–1976) and her second husband, archaeologist Sir Max Mallowan. Christie is best known as the creator of *Miss Marple* and *Hercule Poirot* and for her thriller *The Mousetrap*, the world's longest-running play, which opened in November 1952. She lived at Winterbrook House at Wallingford just to the north-east of the AONB for 40 years.

EAST HENDRED

Heading west along the northern edge of the North Wessex Downs is the spring-line village of East Hendred, the larger of the two Hendred villages; the other being West Hendred. The village was first mentioned in an Anglo-Saxon charter from AD 956, where it was called *Hennerithe*, 'rill of the waterhens'. After the Norman Conquest, the village was divided into five manors, including King's Manor and Abbey Manor.

East Hendred has a wealth of sixteenth- and seventeenth-century brick and timber-framed houses, including a fine example of Tudor brickwork at the village shop.

Hendred House, also known as the Manor of the Arches, is essentially an H-shaped fifteenth-century hall house. Originally owned by the Turbervilles in the mid-twelfth century, the house passed to the Eyston family in the mid-fifteenth century with the marriage of Isabella Stowe (descended from the Arches family) to John Eyston, and has been their family home ever since. The

▲ A great example of Tudor herringbone brickwork in East Hendred

family are related by marriage to Sir Thomas More, one-time Chancellor to Henry VIII: the family still hold More's drinking cup and the staff of John Fisher, Bishop of Rochester; both men were found guilty of treason and executed on Tower Hill in 1535. The adjoining Saxon St Amand's Chapel was founded by Sir John de Turberville after receiving permission from Pope Alexander IV in 1256. The Victorian-Gothic Roman Catholic St Mary's Church, along St Mary's Road, was built by the Eyston family in the 1860s and contains some good stained-glass windows.

The Church of St Augustine of Canterbury has a most interesting, but hidden, faceless clock. Built in 1525 by John Seymour of Wantage (although extensively restored in the 1960s), it is one of the oldest clocks in England, and still chimes the hours and quarters as it has for almost 500 years, while every three hours it plays the *Angel's Hymn* by Orlando Gibbons. The village church was where Prime Minister David Cameron married Samantha Gwendoline Sheffield in 1996.

The little Chapel of Jesus of Bethlehem – now the Champs Chapel Museum – was built by monks from the Carthusian Monastery of Jesus of Bethlehem at Sheen, Surrey, in the fifteenth century on land known as King's Manor – one of the five manors that once made up the village. The manor had belonged to the Abbey of Noyen in Normandy since the time of the Domesday Book until Henry V suppressed foreign religious orders in 1414 and granted the land to the monks from Sheen. The former chapel is now home to the interesting village museum, which tells the story of this once-thriving little medieval market town.

LETCOMBE BASSETT AND LETCOMBE REGIS

Continuing to the west are the neighbouring villages of Letcombe Bassett and Letcombe Regis.

Tucked beneath the downs, Letcombe Bassett has a number of picturesque thatched seventeenth-century cottages. In the Domesday Book it was known as *Ledecumbe*, meaning 'the brook [*lede*] in the valley'; the rest is derived from the Norman baron Richard Bassett, who lived in the manor house in the mid-twelfth century. The mostly thirteenth-century Church of St Michael's and All Angels (partly rebuilt in 1861) has some fine examples of Norman ornamentation. To the left of the entrance is a blocked-up doorway, and still discernible are the four signs of the Evangelists: the Eagle of St John, the Lion of St Mark, the Angel of St Matthew and the Ox of St Luke. Jonathan Swift (1667–1745), Dean of St Patrick's Cathedral in Dublin, satirist, poet and author of *Gulliver's Travels*, stayed at the Old Rectory in 1714.

The 'Regis' of Letcombe Regis was added during the reign of Richard II, although Letcombe was already a Royal Manor to the Kings of Wessex before passing to William the Conqueror. Georgian Letcombe Manor is currently home to the Letcombe Agricultural Research Laboratory. Local gossip has it that a previous owner, an Edwardian lady, was so shocked at two nude statues in the garden that she ordered them to be sunk in the lake: during dredging in 1982 a 6-foot-high second-century Roman marble figure of Hercules was found. St Andrew's Church, which dates from the twelfth century, has the remains of a medieval cross just inside the gate.

WANTAGE

Although located just to the north of the AONB, mention should be made of the market town of Wantage, renamed Alfredston by Thomas Hardy in his novel *Jude the Obscure*, and famed for being the birthplace of Alfred the Great in AD 849. In the Market Place, which has some fine Georgian and Victorian buildings, is a statue of Alfred the Great, King of the West Saxons, who defended his kingdom from the Danes before becoming overlord of England (the statue was erected by Robert Loyd-Lindsay, Lord Wantage – see page 139). Other famous inhabitants have included the poet laureate Sir John Betjeman, who lived here from 1951 to 1972, and the horse racing jockey Lester Piggott who was born here in 1935.

The much-restored church of St Peter and St Paul still displays many features from the original thirteenth-century structure; the restoration was undertaken by well-known Victorian architect George Edmund (G. E.) Street, who lived in Wantage. Inside is a large fourteenth-century altar tomb to Sir William Fitzwaryn (d.1361) and his wife Amica – the Fitzwaryns held the Manor of Wantage from 1207 – and a superb brass figure of Sir Ivo Fitzwaryn (1343–1414), whose daughter Alice married Sir Richard Whittington (1358–1423), four times Lord Mayor of London and the inspiration for the pantomime character Dick Whittington. Opposite the church is the Vale and Downland Museum, dedicated to the geology, history and archaeology of Wantage and the Vale of White Horse.

The former Wilts & Berks Canal, which ran from the Kennet and Avon Canal at Semington to the River Thames at Abingdon, passed just to the north of the town. The 52-mile-long canal opened in 1810 but was abandoned in 1914. At one time Wantage had a short rail link – the Wantage Tramway – to Isambard Kingdom Brunel's famous Great Western Railway that passed by 2 miles to the north. The line opened in 1873, but closed to passengers in 1925 and goods traffic in 1945.

Villages in the Lambourn Valley Area

LAMBOURN

The village of Lambourn, said to have been the model for Thomas Hardy's Maryland in his novel *Jude the Obscure*, is situated in the Lambourn Valley, known as the 'Valley of the Racehorse' owing to the large number of training stables and gallops in the area. The town's name probably means 'the stream where lambs were washed' – 'bourne' is old English for stream, and sheep were important to the area; some of the adjoining names such as Ewe Hill and Sheepdrove reflect this. At the centre of the village is the Market Place, complete with its old stone Market Cross which dates from the reign of Henry VI. Behind the church is the red-brick castellated entrance to the Isbury Almshouses: originally built by John Estbury in 1502, they were rebuilt in 1852 by Henry Hippisley.

The first mention of a church in the village was in Saxon times during the reign of King Alfred in the late ninth century, so there's been a church in Lambourn for over 1,000 years. However, the site may be even older, as the dedication of the church to St Michael the

▲ The castellated entrance to the Isbury Almshouses in Lambourn

Archangel (the 'All Angels' bit was added later) – the destroyer of evil (and therefore paganism) – could indicate that early Christians took over an older site of worship.

The present church has Norman origins, dating from 1180, and is based on a typical cruciform design with a central tower. Inside there are several notable monuments. In St Katherine's Chapel on the north side of the chancel is the impressive alabaster tomb of Sir Thomas Essex of Lambourn Place (d.1558) and his wife, Margaret, while in the Lady Chapel, on the south side of the chancel, there are brass memorials to John de Estbury (d.1372) and his son Thomas; the Estbury family were at one time lords of the manor of Eastbury.

▲ Church of St Michael and All Angels, Lambourn

GREAT SHEFFORD

Travelling south-east along the Lambourn Valley through which the Lambourn Valley Railway once travelled (most of the line closed in 1960) brings us to Great Shefford, which shares part of its name with the much smaller East Shefford (see page 129). The village was mentioned in the Domesday Book as '*Siford*' and the word Shefford is believed to mean 'sheep ford' – a place where sheep could cross the River Lambourn – being derived from the Saxon word '*sciep*', meaning sheep, and ford being a shallow river crossing.

St Mary's Church, which dates from the end of the twelfth century, has a round tower – one of only two in Berkshire, the other being at St Gregory's Church in Welford (see page 124). The font is said to be one of the finest in the county – a superb example of Norman carving. There is also an interesting old preaching cross in the churchyard. The manor house, adjoining the church, was once owned by Sir George Browne (d.1678) who is buried in the churchyard; he was invested as a Knight, Order of the Bath (KB), in 1661.

▲ St Mary's Church in Great Shefford has one of only two round church towers in Berkshire

CHADDLEWORTH

Just to the north-east is Chaddleworth – a place with a London connection. St Andrew's Church, although restored by that great Victorian Gothic revivalist architect George Edmund Street, has a fine twelfth-century Norman doorway and a fairly short thirteenth-century tower. Inside there are a number of memorials along with two 'family pews', or small side chapels. One was built for the Wroughton family – amongst the memorials is the plain wooden cross placed at the grave of Philip Wroughton who was killed at Gaza in Palestine in 1915. The other was built for the Blandy family. A descendant of the family – John Blandy (1783–1855) – emigrated to the Portuguese island of Madeira and founded Blandy's Madeira Wine which celebrated its bicentenary in 2011.

To the south, where RAF Welford air base is now located, was the site of Poughley Priory, although nothing remains today. The priory, which was founded in 1160, was suppressed in 1524 and both Poughley and Chaddleworth were given to Thomas Wolsey for the use of his college – Christchurch – at Oxford. Following Wolsey being charged with treason, his properties were confiscated by the Crown. However, Henry VIII had no need for land in Berkshire so he exchanged it for lands in London. Henry VIII got part of Westminster, including St James' Park, and Westminster gained the Parish of Chaddleworth; the Dean and Chapter of Westminster are still patrons to this day.

Villages to the South of Swindon

CHISELDON AND DRAYCOT FOLIAT

Chiseldon lies adjacent to the A346, just south of junction 15 of the M4. It has a much smaller neighbour Draycot Foliat, although at one time Draycot was the larger village.

Chiseldon, located at the head of a coombe with springs, was first mentioned in AD 903 in a charter of Edward the Elder when the church was granted to the monks at Winchester – this later became Hyde Abbey – and parts of the parish remained with the abbey until the Dissolution. The Church of the Holy Cross was rebuilt by the Normans late in the twelfth century.

Draycot Foliat, like its neighbour Chiseldon, once had its own church, but in 1571 the Bishop of Salisbury ordered the church to be demolished, since neither parish could sustain their own rectors any longer and by that time Chiseldon was the larger village.

In 1881 the railway arrived at Chiseldon with the opening of the Midland and South West Junction Railway; the line closed in 1961 (the disused track-bed now forms the Chiseldon and Marlborough Railway Path which is part of the National Cycle Network). From the outbreak of the First World War in 1914 until 1960 the parish was home to a military camp which had its own railway stop – Chiseldon Camp Halt (see North Wessex at War – pages 51–3). In 2004 the Chiseldon Local History Group opened the Chiseldon Museum which displays a collection of photographs and objects relating to the village over the years; the museum is housed in a redundant chapel in Butts Road. However, one of the most important aspects of the parish was the discovery of twelve Iron Age cooking pots known as the Chiseldon Cauldrons – the largest group of Iron Age cauldrons ever to be discovered in Europe (see page 44).

THE OGBOURNES

Along the River Og, which rises near Draycot Foliat and flows south to join the River Kennet at Marlborough, are the villages of Ogbourne St George and Ogbourne St Andrew along with the hamlet of Ogbourne Maizey; the name Ogbourne is derived from the Saxon 'Oceburnan' or 'Oca's stream'. Ogbourne St George, the larger of the two villages, along with neighbouring Ogbourne St Andrew, has a long history stretching back to Saxon times. In the twelfth century the manorial rights of Great and Little Ogbourne – now Ogbourne St George and Ogbourne St Andrew – were donated by Maud of Wallingford to the Benedictine Abbey of Bec-Hellouin in Normandy and the monks built a priory near to the present Manor House.

Jump forward almost 300 years to the reign of Henry V when all alien orders were suppressed and the estate passed to the Duke of Bedford in 1422. On his death in 1435 the manor passed to the Crown. During the reign of Henry VI the land was granted to King's College in Cambridge, and they held it until 1927. The present church, which stands on the site of the former Saxon church, dates from the twelfth century albeit with extensive Victorian renovations; inside is a brass memorial to Thomas Goddard and his wife, dated 1517.

St Andrew's Church in Ogbourne St Andrew, which dates from the twelfth century with later additions, houses some interesting monuments including two from the seventeenth century to the Goddard family. Within the churchyard are the remains of a fairly large (75 feet in diameter) Bronze Age bowl barrow. Excavations in 1885 by Henry Cunnington located the main Bronze Age cremation, along with an Anglo-Saxon inhumation and around twenty other inhumations that were probably part of the medieval churchyard. A local legend mentions that the mound was the 'abode of venomous vipers'.

BROAD HINTON

To the west of Chiseldon is the village of Broad Hinton, whose name (Hinton) means 'high farm or town'. At one time it was known as Hinton Wase after the Wase family who owned the manor. Other owners over the centuries have included the Wroughtons, followed by the Glanvilles and even the Duke of Wellington.

Although there may well have been an earlier church here, the present Church of St Peter ad Vincula (or St Peter in Chains) – one of only around fifteen churches in England with this dedication – dates from the late twelfth century. The perpendicular styled tower was added in the fifteenth century and the late-nineteenth century restorations in a Gothic Revival style were undertaken by C. E. Ponting (1850–1932). Inside there are several monuments to the Wroughtons and Glanvilles.

One particular monument in the nave, dated 1597, is to Sir Thomas Wroughton. Sir Thomas's effigy, along with those of his eight children, is without hands, unlike his wife, Anne, who is holding an open Bible. Maybe the hands have been broken off during the lifetime of the monument, but the fact that they are missing gave rise to a local legend. It is said that Sir Thomas came home one day to find his wife reading the Bible instead of cooking his meal. He threw the Bible in the fire, but she retrieved it, badly burning her hands. Due to his blasphemous behaviour, Sir Thomas's hands and those of his children are said to have withered away.

CLYFFE PYPARD

Further west, on the lower edge of the steep escarpment as it drops away to the much flatter ground of the Wiltshire plain, is the village of Clyffe Pypard, named after the thirteenth-century lord of the manor, Richard Pipard, and its location on the steep slope or 'cliff'.

Records show that there was a church here in the thirteenth century, though the present St Peter's Church dates from the fifteenth century and has undergone extensive nineteenth-century renovations. Inside are a wagon roof, an ornate early seventeenth-century pulpit and several memorials including a striking white-marble effigy of Thomas Spackman, a local carpenter who died in 1786, leaving a large sum of money in his will to pay for the education of the village's poorer children. There is also a memorial brass to Henry de Cobham (d.1380) who was at one time lord of the manor. In the churchyard is the grave of Sir Nikolaus Pevsner (1902–1983), author of the county-by-county series of books on the *Buildings of England*, and his wife.

In 1530 William Dauntsey sold the manor to John Goddard of Upper Upham House near Aldbourne (d.1542) and the family then held the manor for several centuries before it passed by marriage to William Wilson in the twentieth century. The village pub, The Goddard Arms, displays the three-crescent Goddard coat of arms along with the family motto *Cervus non servus* or 'the stag is not a slave'.

The Vale of Pewsey and the River Bourne Valley

PEWSEY

Right at the heart of the Vale of Pewsey, aptly described by William Cobbett in his *Rural Rides*: 'I never before saw anything to please me like this valley of the Avon', to the south-west of Marlborough, is the village of

Pewsey, from which the vale takes its name. To some it may look like a small town in terms of size, but Pewsey was never granted a charter to hold a market, so village it is.

Pewsey's own history can be traced back to Saxon times, but we know from archaeological finds that the area's history stretches right back to the Neolithic period. At one time Pewsey was held by one of the greatest Saxons of all – Alfred the Great – who was crowned King of Wessex in AD 870; at this time it was known as *Pevisigge*. In AD 940 Alfred's grandson, Edmund, granted the royal estate to St Peter's Abbey in Winchester, later known as Hyde Abbey. By the time of the Domesday Book the village was known as *Pevesie*, and the abbey continued to hold the manor until the time of Henry VIII's Dissolution of the Monasteries.

The manor was granted to Edward Seymour, 1st Duke of Somerset in 1547, before passing to the Crown on his execution in 1552 and then back into the hands of his son, Sir Edward (d.1621). The manor then passed through several Dukes of Somerset (the Seymours also held Tottenham House at Savernake) before being sold to Joseph Champion in 1776, who then sold it on to St Thomas's Hospital, London, in 1770. It was around this time that the village rector was one Joseph Townsend (1739–1816); in fact he was rector for over fifty years from 1764 until his death – and was rather a special man.

Having studied medicine he invented 'Townsend's Mixture' – a concoction of mercury and potassium iodide – as a treatment for syphilis; he ran a free medical service for the poor; he defended the biblical creation story in *The Character of Moses as an Historian, Recording Events from the Creation to the Deluge* (1813); wrote about his travels through Spain in *A Journey Through Spain in the Years 1786 and 1787*; helped William Smith (who created the first nationwide geological map and was known as the 'Father of English Geology') on the chronological sequence of geological strata; and within the village he helped improve the roads, including building a bridge across the River Avon in 1797 connecting the High Street with River Street.

Major transport improvements came to Pewsey with the opening of the Kennet and Avon Canal in 1810, quickly followed by the arrival of the railway with the opening of the Berks & Hants Extension Railway in 1862; this now forms the line from Reading to the West Country.

In the centre of the village, at the junction of the High Street (B3087) and the A345, is the Market Place – a fairly recent naming, probably within 200 years, rather than relating to a medieval market. Here, a statue of Alfred the Great was unveiled in 1913 to commemorate the coronation of King George V.

Pewsey has many interesting buildings, such as the thatched and timber cruck-framed Ball House, located on Ball Corner, which dates back to the fourteenth century and is probably one of the oldest houses in the village. Others include the Court House by the church and Bridge Cottage on the High Street. Part of the old iron foundry just off the High Street, established by George Whatley, is now home to the Pewsey Heritage Centre and houses an interesting collection of artefacts covering aspects of life in the village over the preceding 150 years.

Pewsey's first church, which was replaced by a Norman structure, was Saxon and was probably made of wood, with a thatched roof. Parts of the Norman building still exist in the present church of St John the Baptist. During the fourteenth century the nave roof was rebuilt with the addition of a clerestory (a row of windows high up in the church that allow more daylight into the building) and the tower was added in the sixteenth century; much later the building underwent major renovations at the hands of Victorians including C. E. Ponting from nearby Collingbourne Ducis.

Inside, atop the Norman font, is a large elaborate font cover carved by the Revd Bertrand Pleydell-Bouverie (1849–1926) who was rector from 1880 to 1909; incidentally he was the second Pleydell-Bouverie to be rector, the first being the Revd Frederick Pleydell-Bouverie (1785–1857). The Revd Bertrand was also responsible for the carved altar rails, said to have come originally from a Spanish battleship – the 112-gun *San Josef*, captured by Lord Nelson off Cape St Vincent in 1797.

The Pewsey Carnival, a tradition that still flourishes

today, was first held in 1898 and includes many events such as The Feaste. The origins of the feast day are claimed to date back to the time of Alfred the Great when, following a victory over the Danes, he stood at the church and announced that the day – 14 September, Holy Cross Day – should be a feast day forever.

Over the years Pewsey has had two carved white horses, although the first one has long since disappeared (for information about Pewsey's white horses, see page 142).

VILLAGES OF THE RIVER BOURNE VALLEY

Along the River Bourne as it travels south from its source towards the River Avon are the villages of Burbage, Collingbourne Kingston and Collingbourne Ducis.

Burbage is situated on a watershed at the eastern end of the Vale of Pewsey; the streams to the east drain into the River Dun and then on via the River Kennet to the River Thames; streams to the west head for the River Avon, while the River Bourne drains south to later join the River Avon. All Saints Church dates back to at least the early thirteenth century, although it was largely rebuilt in the nineteenth century. The Kennet and Avon Canal to the north of the village opened in 1810 and this was followed by the Great Western Railway's Berks & Hants Extension Railway from Hungerford to Pewsey and beyond. The line which follows the canal to the north of the village opened in the 1860s and now forms part of the line from Reading to Westbury.

Next stop heading downstream is Collingbourne Kingston. The name Collingbourne, which is shared with its close neighbour Collingbourne Ducis, dates back to Anglo-Saxon times, being derived from 'the stream of Cola's people' – the stream in question being the River Bourne. The manor was held by St Peter's Abbey in Winchester, later known as Hyde Abbey, and because of their ownership, the manor was known as Collingbourne Abbots. Following the Dissolution of the Monasteries, Henry VIII granted the manor to Edward Seymour, brother of Henry's third wife, Jane Seymour, in 1544. The manor then descended from the Seymours through the Bruce, Brudenell and Bruce-Brudenell families until most of the estate was sold in 1929.

The village church dates back to the late eleventh or early twelfth century and was originally dedicated to St John the Baptist before being renamed St Mary's in 1344.

The Swindon, Marlborough & Andover Railway, which opened in 1882 and later became part of the Midland & South Western Junction Railway, followed the River Bourne south to the village of Collingbourne Ducis; the line closed in 1961.

In the thirteenth century Collingbourne Ducis was known as Collingbourne Earls, after the lord of the manor, the Earl of Leicester. Sometime later John of Gaunt inherited the manor and, on his becoming Duke of Lancaster, the village became known as Collingbourne Ducis or 'Dukes'. The Gothic Revival architect C. E. Ponting (1850–1932) was born in the village and the restoration of St Andrew's Church in 1856 by George Edmund Street is said to have made a lasting impression on him.

Along the North Hampshire Downs

GREAT BEDWYN

In Saxon times Great Bedwyn was known as *Bedanheaford* – meaning 'the grave's head' – and was the residence of the Saxon chief Cissa. By the time of the Domesday Book the town was known as *Bedewinde* and was held by the king.

The village, located to the south-west of Hungerford on the banks of the River Dun, has both the Kennet and Avon Canal and the railway from Reading to Westbury passing through it. The canal arrived in 1810 followed some years later with the opening of a line from Hungerford in 1862; this was an extension to the GWR-backed Berks & Hants Railway that operated between Reading and Hungerford and had opened in 1847. On the wall of the post office, close to the church, is a selection of works that were undertaken at Lloyd's stonemason's yard which was established in 1790; sadly the yard closed in 2009.

TOWNS AND VILLAGES

▲ Church of St Mary the Virgin, Great Bedwyn

The Church of St Mary the Virgin dates from 1092, though most of what is visible nowadays is from the twelfth and thirteenth centuries; it is one of the largest churches in the area. Inside is an impressive monument to Sir John Seymour, father of Jane Seymour who married King Henry VIII in 1536, becoming his third wife. She died in 1537 shortly after the birth of their son, who later became Edward VI. The church also holds the stone figure of a knight, believed to be Sir Adam de Stokke (d.1313), and the tomb of Sir Roger de Stokke (d.1333), son of Sir Adam.

Just to the north of the village are the remains of a former Iron Age hill fort at Chisbury and, sited on the edge of the earthworks, adjacent to Chisbury Manor Farm, is the empty shell of St Martin's Chapel. Although used as a barn for the last 300 years, the medieval chapel dating from the thirteenth century, with flint walls and a thatched roof, was originally constructed by the lord of Chisbury Manor as a chapel-of-ease so that he, and local people, did not have to travel to Great Bedwyn. The building ceased to be used as a consecrated chapel in 1547 and is now in the guardianship of English Heritage (accessible from the road between Chisbury and Little Bedwyn).

SHALBOURNE

To the south-west of Hungerford, tucked below the steep scarp slope of the North Hampshire Downs, is Shalbourne, first recorded in the tenth-century Anglo-Saxon charters as *Scaldeburnam*, derived from '*scealde burna*' meaning shallow stream.

▲ The former St Martin's Chapel at Chisbury

Shalbourne Manor, once known as 'Eastcourt', passed by marriage to Robert de Tatteshall (c.1241) and became known as 'Tateshale'. About 300 years later the manor was sold to Edward, Duke of Somerset, and stayed with the family until it passed by marriage to the Earl of Ailesbury; the estate then descended with the family until most of it was sold in 1929.

The Church of St Michael and All Angels dates from the twelfth century, with a fifteenth-century tower. The church has five bells dating from the mid-seventeenth century, which have inscribed makers' names: Henry Knight of Reading; Cor of Aldbourne; and Robert Wells of Aldbourne. A sixth bell made by the Whitechapel foundry was added in 1983. Memorials inside include a sixteenth-century Elizabethan monument to Sir Francis Choke and a tablet to the agriculturalist Jethro Tull, who invented the forerunner to the modern seed drill and took over his great-uncle's farm – Prosperous – from 1709 until his death in 1741.

The village still has interesting agricultural links. Charles Flower acquired Carvers Hill Farm in 1980 with the aim of demonstrating practical methods of countryside restoration within the framework of everyday farming practices. The farm, situated on the edge of the village, includes chalk, greensand and clay soil types, and is now a mixed farm with corn crops, pasture and wildflower meadows. The seeds from the wildflower meadows are harvested and sold to both individuals and local authorities to help improve the countryside for insects and butterflies, which in turn helps the local bird populations.

Although Shalbourne's mill was derelict by the late sixteenth century, it had been fully rebuilt a hundred years later and remained in operation until the 1920s. Following the closure of the mill, watercress beds were established along the Shalbourne brook, which is a tributary of the River Dun. Varieties such as 'Morning Dew' and 'Mill Brand' were sent to markets at Covent Garden and Birmingham; sadly the business closed in 1972.

The artist Eric Ennion (1900–1981), who from an early age was fascinated by watching and drawing birds, moved to Shalbourne's Mill House in 1961, from where he continued to paint and run courses on wildlife and landscape painting until his death.

BUTTERMERE

A short distance to the south-east of Shalbourne, situated adjacent to the county boundary with neighbouring Hampshire and Berkshire, is Buttermere. At 830 feet above sea level the little village is the highest in Wiltshire, with a church – St James the Great – reputed to be the county's smallest. The church was standing in the thirteenth century, though it was largely rebuilt in the mid-nineteenth century.

FACCOMBE AND ASHMANSWORTH

Tucked in the far north-west corner of Hampshire, close to Pilot Hill which, at 938 feet, is the highest point in the county, are the neighbouring villages of Faccombe and Ashmansworth.

For many years the main village in the area was Netherton. However, within the last 200 years, the population has drifted up to Faccombe, located less than a mile to the east, and just to the south of Pilot Hill. St Barnabas' Church at Faccombe, built in the second half of the nineteenth century, has a decorated Norman font and several seventeenth-century memorials from Netherton's original thirteenth-century church.

Slightly further south-east is Ashmansworth, Hampshire's highest village, with a lovely collection of thatched cottages. Back in Saxon times the manor of Ashmansworth was granted to the Church of Winchester before passing to the Bishop of Winchester; the manor remained with the bishopric until 1649 when it was sold to Obadiah Sedgwick, a minister in London's Covent Garden. Yet by 1660 it was back in the hands of the bishopric until the start of the nineteenth century, when it passed to the Herbert family and the Earls of Carnarvon.

The twelfth-century St James' Church, originally dedicated to St Nicholas, is well worth a look. Inside are fragments of medieval wall paintings, while in the porch are two engraved windows. The English composer Gerald Finzi (1901–1956) lived opposite the church from 1939 until his death in 1956, and in the 1970s his

▲ Engraved glass window in St James' Church celebrating fifty English composers including Gerald Finzi

wife, Joy, raised the funds to have a window designed by Sir Laurence Whistler and engraved by his son, Simon, in celebration of English composers. The window portrays the 'tree of music', with musical notes as leaves and the initials of fifty composers, along with their year of birth, attached to the roots. The second window, again designed by Sir Laurence Whistler, is in memory of Joy Finzi.

SYDMONTON

Further east along the scarp is Watership Down, forever linked with a group of rabbits in Richard Adams's book of the same name. Tucked below is Sydmonton Court, the country home of composer Lord (Andrew) Lloyd-Webber of Sydmonton, writer of such great musicals as *Joseph and the Amazing Technicolor Dreamcoat*, *Starlight Express*, *Phantom of the Opera*, *Cats* and *Evita*.

Although of Tudor origin, the house has seen many additions and alterations over the intervening centuries. The estate, which may be named after Sydeham who was mentioned in a Saxon charter of AD 931 relating to land at nearby Ecchinswell, was owned by Romsey Abbey until Henry VIII's Dissolution of the Monasteries. Sydmonton was then granted to John Kingsmill of Whitchurch (d.1556) and stayed in the family for many generations. In the nearby Church of St Mary in Kingsclere is a large altar tomb to Sir Henry Kingsmill (d.1625) and his wife, Bridget White (d.1672). On the slab are two recumbent alabaster effigies; Sir Henry is in the armour of the time of Charles I and his wife wears a veil and a tight bodice.

Hampshire Villages in the Bourne Rivulet Valley

The villages of Hurstbourne Tarrant, St Mary Bourne and Hurstbourne Priors lie alongside the Bourne Rivulet – often just called 'The Bourne' by locals – which forms a tributary of the River Test. This river, the upper reaches of which are seasonal, only flowing during the wetter months when the water level in the chalk aquifer is high, was described by the Irish baritone singer and passionate fly-fisherman Harry Plunket Greene (1865–1936) as 'unquestionably the finest trout stream in the south of England' in his book *Where Bright Waters Meet* (1924); Plunket Greene rented a house in Hurstbourne Priors for ten years at the start of the twentieth century.

HURSTBOURNE TARRANT

The 'Tarrant' part of Hurstbourne Tarrant originates from the early thirteenth century, when the village was given to the Cistercian nunnery of Tarrant Kaines. St Peter's Church dates from the late twelfth century and the south doorway is a fine example of Norman workmanship. The American-born artist and writer Anna Lea Merritt (1844–1930), famed for her nude painting *Love Locked Out* (1890), lived in the village from 1891 until her death. In 1902 she wrote and illustrated a book, *A Hamlet in Old Hampshire*, giving a portrait of Hurstbourne Tarrant.

▲ Picturesque thatched cottage beside St Peter's Church in St Mary Bourne

During the Second World War land close to the village was used as a decoy site for RAF Andover, the headquarters of RAF Maintenance Command. The decoy site, designed to deceive enemy aircraft into attacking a spurious target, included fake aircraft and buildings.

ST MARY BOURNE

Slightly further downstream from Hurstbourne Tarrant is the picturesque village of St Mary Bourne. There are features inside the twelfth-century St Peter's Church that make it well worth a visit. The stained-glass windows include one in the north aisle given by the Bank of England – this window contains the figure of St Christopher surmounted by the 'Old Lady of Threadneedle Street'. Along the south wall is the effigy of Sir Roger des Andelys of Wyke Manor, who fought in the Crusades. Perhaps the most striking feature is the beautiful twelfth-century square, polished black 'marble' font (although the material is often called

▲ A rare Tournai font inside St Peter's Church, St Mary Bourne

'marble' it is actually a form of black limestone), one of only seven in England, four of which are located in Hampshire. This was made at Tournai in Belgium, close to the border with France. Two sides of the font are carved with Norman arches, one side being surmounted by four doves and the other with fleur-de-lys; the remaining two sides display stylized vines.

HURSTBOURNE PRIORS

Continuing downstream, passing under the imposing nineteenth-century brick viaduct which carries the West of England railway line across the Bourne Rivulet, is Hurstbourne Priors, right on the southern edge of the AONB.

The first mention of Hurstbourne Priors, or *Hissaburna*, was in AD 790 when the land was granted by Beorhtric (King of Wessex between AD 786 and 802) to Prince Hemele in exchange for land on the River Meon; the prince then granted the manor to the monks of Abingdon. By the time of the Domesday Survey the manor was held by the Church of Winchester and they continued to hold the manor until the Dissolution of the Monasteries, when it reverted to Henry VIII. The manor then passed through several owners until 1558 when Sir Robert Oxenbridge (d.1574) bought it and the family continued there until 1636, when it was sold to Sir Henry Wallop of Farleigh Wallop (d.1599); the manor then remained with the Wallops for 300 years. A later member of the family – John Wallop (1690–1762) – was created Viscount Lymington in 1717 and then 1st Earl of Portsmouth in 1743.

More recently the prominent newspaper magnate, Lord Camrose (1879–1954), acquired the estate.

The village had a church as far back as AD 820 when it was mentioned in a charter of Denewulf, Saxon Bishop of Winchester. The present Church of St Andrew the Apostle, which stands on the site of the original Saxon church, was built by the Normans in the twelfth century. The north chapel was added in the sixteenth century and in the eighteenth century the south transept, or Portsmouth Aisle, was built for the use of the Earl of Portsmouth (lord of the manor) and his family. In the north wall of the chancel is the sixteenth-century canopied Oxenbridge Altar Tomb, complete with the recumbent figures of Sir Robert Oxenbridge (d.1574), one time lord of the manor, and his wife. Their children are depicted along the side of the tomb, some holding skulls to indicate that they predeceased their parents.

Behind the church, close to the river, is a rather large horse chestnut tree, claimed to be one of the largest such trees in England, with a height of 115 feet and a girth of just over 23 feet.

At one time the village lay close to three railway lines, although only one – the main West of England line running from London Waterloo to Exeter – is still open; unfortunately the local station closed in 1964. Just to the east, passing through Whitchurch, was the Didcot, Newbury & Southampton Railway (DN&SR). The southern section between Newbury and Winchester opened in 1885, with the final section to Southampton opening in 1891; the line closed in the early 1960s. The competing London & South Western Railway (LSWR) opened a branch line, the Fullerton Junction to Hurstbourne Line, in the 1880s; this line ran between Hurstbourne Priors, which was on the route passing through Basingstoke and Salisbury, and Fullerton to the south of Andover, where it connected with a line heading towards Southampton. Passenger services were withdrawn in 1931, although the line stayed open for freight traffic before finally closing in the 1950s.

CHAPTER FIVE

From stately homes to white horses

Throughout history, our ancestors have left their mark on the landscape. We looked at the prehistoric remains in Archaeology Uncovered in Chapter 2; now let us take a virtual tour through some of the more recent stately homes, picturesque churches, monuments and hill carvings that lie within the North Wessex Downs (grid references are given to allow easy location on maps).

Stately Homes

Within the North Wessex Downs there are a number of stately homes; however, only those that are either open to the public, or visible from public rights of way, are described here.

ARDINGTON HOUSE

Overlooking a small lake in the Oxfordshire village of Ardington is the beautifully symmetrical grey and red-brick Baroque-style Ardington House (SU432883). The house, built for Edward Clarke in 1720 (around the start of the Georgian period) and once owned by Lord Wantage, has been the Baring family home for several generations. The Barings – at one time wool merchants in Exeter – founded Barings Merchant Bank, which hit the world headlines when it was brought down by the unauthorized trading activities of Nick Leeson in the 1990s.

▲ Picturesque Ardington House

▲ Marble figure by Edward Bailey inside Holy Trinity Church, Ardington

▲ Ashdown House – the 'perfect doll's house' – tucked into the Lambourn Downs

Hidden within the house is a stunning imperial staircase with two matching flights leading into one – considered by experts to be one of the finest examples in Britain – and a wood-panelled dining room said by the former poet laureate Sir John Betjeman to be 'the nicest dining room in Oxfordshire'.

Nearby is the Holy Trinity Church which dates from around 1200 and has a fine Norman doorway and a fourteenth-century churchyard cross, one of only a few in the county. Inside there is a marble figure by Edward Bailey, who also designed Nelson's Column in Trafalgar Square, and memorials to the Clarke family who have held the manor here for nearly 500 years.

ASHDOWN HOUSE

Located high on the Lambourn Downs between the Ridgeway and the village of Lambourn, the Ashdown Estate has a considerable history, having been mentioned in an Anglo-Saxon Charter dated AD 947, when King Eadred of the West Saxons granted land at '*Ayssehudun*' to Edric, one of his thanes or *thegns* (a 'king's follower'). For several hundred years the estate was held by Glastonbury Abbey until Henry VIII's Dissolution of the Monasteries. Later, in 1625, the estate was bought by the Craven family.

Located on the estate is Ashdown House (SU282820), an unusual Dutch and French-styled building, described by the architectural historian Nikolaus Pevsner as 'the perfect doll's house'. The four-storey building crowned with a cupola, constructed of locally quarried chalk with Bath stone quoins, string courses and moulded windows, was commissioned by William Craven, 1st Earl of Craven (1608–1697), in the early 1660s as a hunting lodge and house fit for the queen he loved, Queen Elizabeth of Bohemia.

William Craven, a courageous man loyal to the Stuart cause, was one of the richest men in England at the time, having inherited his money from his father, and he owned large amounts of land including the Ashdown Estate, Coombe Abbey and Hamstead Marshall (see later in this chapter). During the English Civil War (1642–51) Craven gave King Charles money to help the Royalists fight against the Parliamentarians and a grateful King made him an Earl.

Elizabeth Stuart, the daughter of James I of England and VI of Scotland (thus sister of Charles I), had married Frederick, the Elector Palatine, and in 1619 they reigned as King and Queen of Bohemia for just one year before Frederick was defeated at the Battle of White Mountain. Following the defeat they were forced to live in exile in The Hague and Elizabeth was forever-after known as the Winter Queen.

After Frederick's death, Craven, who had first met Elizabeth when he was a young soldier, provided financial support to Elizabeth and when she returned to England she lived in his house in Drury Lane. However, Craven was concerned about the plague in London and wanted to build Elizabeth a fitting home in the relative safety of the country. Knowing her love of hunting he chose Ashdown. Unfortunately, Elizabeth died in 1662 without ever seeing the house.

On William's death, with no direct heir, the earldom became extinct and all of his estates passed to another William Craven, the grandson of the Earl's cousin, and he became the 2nd Baron Craven (1668–1711).

The Craven family were fond of hunting and horses and both the 3rd Baron Craven (1700–1739) and the 4th Baron Craven (1702–1764) held races and hunts on their land – see the Galloping Horses section on page 63 for more information.

Throughout the eighteenth century the Lords Craven lived mostly at one of their other properties, either Hamstead Marshall or Coombe Abbey, using Ashdown only for sport. Then William, 7th Baron Craven (1770–1825), was created the 1st Earl of Craven (second time round) by George III in 1801 in recognition of his services to the Crown. His son inherited Ashdown House in 1825 and throughout the rest of the nineteenth century the house was used as a family home. The family's horse racing heritage continued with the 2nd Earl who also established a cricket club at Ashdown around 1840, and a golf course to the west of the house.

From 1924 the house was let to tenants and during the Second World War it was occupied by troops from both the UK and America. Sadly, at the end of hostilities, the house was left in a near-derelict state, although it remained with the Craven family until it was given to the National Trust by Cornelia, Dowager Countess of Craven, in 1956. More recently the guitarist and songwriter for the rock group The Who, Pete Townshend, bought a 40-year lease on the house in 2010.

Lying in a field just to the east of the house are a large number of sarsen stones and the field has been designated a Site of Special Scientific Interest (SSSI) owing to the population of lichens that live on the stones. To the north-west of the house are the remains of an Iron Age settlement known as Alfred's Castle (see Archaeology Uncovered – page 43).

AVEBURY MANOR

Mention Avebury and most people will think of the famous Avebury henge and stone circle, yet the Wiltshire village is also home to a manor house (SU098700) that dates back to the sixteenth century.

Over the centuries the manor has had many owners who have left their mark on the building, including the archaeologist Alexander Keiller who undertook numerous excavations at nearby prehistoric sites in the 1930s.

The manor house sits on the site of a Benedictine 'alien' priory that had been founded in the early twelfth century as a cell of the monastery of Saint Georges de

▲ Avebury Manor House – once the home of Alexander Keiller

Boscherville near Rouen in Normandy. During the Hundred Years' War with France, Henry V had all alien (foreign) priories dissolved and its property was granted to Fotheringhay College; the college then held the manor until the Dissolution of the Monasteries in the sixteenth century.

The present manor house was built by William Dunch between 1555 and 1580, although it was greatly enlarged around 1601 by Sir James Mervyn. The rooms which best represent this phase of the house's history are the Great Parlour and Tudor Bedroom, which both retain their fine ribbed plaster ceilings and carved stone fireplaces. Around 1740 Richard Holford undertook remodelling of the Great Hall to bring it in line with the latest fashion. The last major phase of alteration of the house was carried out by Colonel and Mrs Jenner who first tenanted and then owned the house in the early twentieth century, and were keen to restore many of the rooms to their late sixteenth- and early seventeenth-century style; the National Trust acquired the manor house in the 1990s.

During 2011 the manor house underwent a full refurbishment for a BBC TV programme, *The Manor Reborn*, marking another chapter in the house's history. This groundbreaking project brought the manor, which had been empty since 2009, back to life in such a way that the visitor can embark on a journey through time, admiring rooms decorated in varying styles to reflect some of the people who lived at, or visited, the manor throughout its history. These range from a Tudor marriage room inspired by the marriage of widowed Debora Dunch to the High Sheriff of Wiltshire in the late 1590s, to a Queen Anne bedroom recreated using brightly coloured panels representing different marbles; from the Georgian-period room reflecting the time of the Williamsons, where the walls are covered with hand-painted wallpaper from China, to the billiard room depicting the twentieth century, where visitors can sit and read a book – or even play billiards. The little parlour at the back of the house evokes the time of archaeologist Alexander Keiller, ready and waiting to host a 1930s evening cocktail party for his field staff.

BASILDON HOUSE

In Basildon House (SU611781) we have a beautiful example of a Palladian mansion, built between 1776 and 1783 by John Carr of York for Sir Francis Sykes (1730–1804), 1st Baronet, who had made his fortune with the East India Company. Little is known about the early years of the Basildon Estate until it was acquired by the Fane family in the mid-seventeenth century and remained with them until it was bought by Sir Francis in 1771 from the family of the 2nd Viscount Fane. The old house was demolished to make way for his new mansion, with its honey-coloured Bath stone west façade and upper floor entrance stairs hidden behind an impressive portico. Unfortunately for Sir Francis he passed away before it was completed; he is buried at the nearby St Bartholomew's Church in Lower Basildon.

The estate passed to his son, the 2nd Baronet, who unfortunately died a few months later, the estate being inherited by his young son, Sir Francis William Sykes (1799–1843), 3rd Baronet. By this time the family finances were starting to suffer and Sir Francis later squandered his grandfather's remaining fortune. To add to his woes, his wife had an affair with the painter Daniel Maclise (1806–1870), who was a great friend of Charles Dickens. Sykes publicly denounced the painter, causing a high society scandal that is claimed to have resulted in Charles Dickens naming his villainous character Bill Sikes, in *Oliver Twist*, after Sir Francis by way of revenge.

In 1838 the house was sold to Liberal MP James Morrison (1789–1857) and it was during his ownership that the mansion was finally completed. Morrison, the son of an innkeeper from Hampshire, married Mary Anne, daughter of Joseph Todd, a London draper, and he quickly made it one of the largest drapery businesses in the world. Following his death, the estate passed to his daughter, Ellen, until her death in 1910, when it was inherited by a nephew, James Morrison. In 1914, the house was requisitioned by the British Government and used as a convalescent home for injured soldiers and, in 1929, James was forced to sell the estate.

The house saw further military use during the Second World War and was facing demolition when, in 1952, the newspaper magnate Lord Iliffe (Edward Langton Iliffe, 1908–1996) and his wife, Lady Iliffe, stumbled across it; she remarked at the time: 'To say it was derelict, is hardly good enough, no window was left intact and most were repaired with cardboard or plywood.' They lovingly restored the house to its former grandeur and in 1977 it was donated to the National Trust.

The house was used as the location for Netherfield Park in the 2005 film of Jane Austen's *Pride and Prejudice* and as Lord and Lady Radley's house in *Dorian Gray* (2009).

ENGLEFIELD HOUSE

Englefield House, situated just to the north of Theale (SU622719), is part of the large Englefield Estate owned by the Benyon family for more than 250 years, although the estate was first mentioned over 1,000 years ago. The name Englefield may be derived from a battle that was fought here between the Saxons (under Ethelred of Wessex) and the Danes in AD 871 and means 'Englishmen's (battle) field'.

Throughout the Middle Ages, the manor was held by the Englefield family until the estate was confiscated from Sir Francis Englefield, a Catholic, in 1559 by Elizabeth I after he had fled to Spain to escape persecution following the rise of the Protestant religion. The house was then granted to Sir Francis Walsingham – Elizabeth's spymaster-general – and after a succession of short-lived residents, the estate was purchased by John Paulet, 5th Marquess of Winchester, known as the 'loyal marquess' for his defence of Basing House in Hampshire during the Civil War. He married Honora, granddaughter of Sir Francis Walsingham, and it is through this family that Englefield passed to the Benyons.

In the late nineteenth century, Richard Fellowes Benyon rebuilt the villagers' houses as a model estate village and provided them with such modern amenities as a bathing pool, soup kitchen and a new school; he also modernized the nearby church.

The earliest part of St Mark's Church dates from late twelfth century; however, it was extensively restored by George Gilbert Scott in 1857 and the Early English-style tower (i.e. a facsimile of thirteenth-century style) was added in 1868. Inside there are two effigies believed to be members of the Englefield family: one is a late thirteenth- or early fourteenth-century effigy of a cross-legged Crusader knight in full armour, with feet resting on a lion and in the act of drawing a sword; the other is an effigy of a woman dating from about 1340. Other memorials include those to John Paulet, 5th Marquess of Winchester, and his wife Honora who died in 1660.

For movie buffs, Englefield House has been used in several films, including *Easy Virtue* (2008), a romantic comedy based on a play by Noel Coward; *X-Men: First Class* (2011), where the house doubles as the X-Mansion – the home of Charles Xavier and his training centre for mutant youths; *The King's Speech* (2012), based on the true story of King George VI; and *Great Expectations* (2012), where it stood in for *Satis House*, home of Miss Havisham who withdrew from the world after being jilted on her wedding day.

HAMSTEAD MARSHALL HOUSE

Hamstead Marshall, lying close to the banks of the River Kennet (SU421667), dates back to at least Saxon times, although the first written evidence appears in the Domesday Book of 1086, when it was held by a Norman called Hugolin the Steersman. Hamstead's importance started to grow in the early twelfth century when it became the seat of John Marshal (d.1165) who was given the hereditary title of Earl Marshal, one of the King's important military officers. The manor and title later passed to his son, also called John, and on his death to William (1147–1219); William Marshal, who served several English kings, was described by the Archbishop of Canterbury (Stephen Langton, d.1228) as the 'greatest knight that ever lived' (he later became the 1st Earl of Pembroke). William was buried at Temple Church, London, which was built by the Knights Templar.

During the sixteenth century, Queen Elizabeth I gave the manor to Sir Thomas Parry by way of a reward for his loyal service while she had been kept a virtual prisoner by her sister, Queen Mary. Sir Thomas' son had the original house pulled down to make way for a fine Tudor mansion; however, this was ruined by Parliamentary troops stationed in the park during the 1st Battle of Newbury.

▲ One of the pairs of ornate brick gateposts, all that remains of the once grand Hamstead Marshall House

▲ Another of the gateposts from the former Hamstead Marshall House

▲ Highclere Castle: home of Lord and Lady Carnarvon and the setting for the TV series *Downton Abbey*

In the early seventeenth century the manor was acquired by the Craven family and their ownership of the estate continued into the twentieth century. The 1st Earl of Craven, who loved Elizabeth, one-time Queen of Bohemia and sister of Charles I (see Ashdown House, earlier in this chapter), commissioned the Dutch architect Sir Balthazar Gerbier to design a smaller version of her former home, Heidelberg Castle in Germany. The adjacent St Mary's Church, parts of which date from the twelfth century, with later medieval additions and a fine red-brick Berkshire tower added around 1622, houses a memorial to the architect.

Sadly, Elizabeth died before construction work had even started; however, the Earl continued with the building as a monument to her memory. Unfortunately the house burnt to the ground in 1718 and all that is left of the once-vast mansion are eight pairs of elaborate gateposts standing in the open fields.

The present Hamstead Lodge, based closer to the canal, probably dates from 1720, though later modified to a Regency style. During the nineteenth and twentieth centuries several of the Craven title-holders died young and death duties took a heavy toll on the family fortune. In 1967 Hamstead Lodge was leased as a nursing home and in the following decade many smaller Craven-owned properties were auctioned off. In 1984 the remainder of the estate was sold and the new owner converted the lodge back to a private house once the nursing home lease had expired.

HIGHCLERE CASTLE

Lying just to the south of Newbury where the Hampshire Downs rise up from the Kennet Valley is Highclere Castle (SU445587), a truly stunning early-Victorian mansion.

The Highclere Estate, which originally belonged to the Bishops of Winchester for 800 years, was purchased by Sir Robert Sawyer, Attorney General to Charles II and James II in 1679, and he bequeathed the house and estate to his only daughter, Margaret, in 1692. Her subsequent marriage to the 8th Earl of Pembroke brought Highclere into the Herbert family, ancestors of the Earls of Carnarvon.

In 1838 Sir Charles Barry (1795–1860) – architect of the Houses of Parliament – was asked by Henry Herbert, 3rd Earl of Carnarvon (1800–1849), to design him a new house – although on two occasions Barry's designs were rejected for being too modest, since the Earl wanted a grand mansion that would impress the world. The house was not rebuilt, but rather the original brick Georgian house was encased in an exuberant Victorian neo-Gothic stone façade. Highclere was finally completed after both the Earl and Barry were dead; the 4th Earl (1831–1890) was responsible for the sumptuous interiors.

Close to the present 'castle' are the ruins of the parish church built by Sir Robert Sawyer around the time of the Elizabethan house, which was later replaced by both the Georgian mansion and then the present Victorian structure.

▲ The Temple of Diana, one of several follies within the grounds of Highclere Castle

The extensive parkland covering 1,000 acres, designed by the acclaimed eighteenth-century landscape gardener Lancelot 'Capability' Brown, includes several follies to improve the vista, such as the pillared temple called Jackdaws Castle. Built in 1743 by Robert Herbert (whose nephew, Henry Herbert, became the 1st Earl of Carnarvon in 1793), the folly incorporates Corinthian columns salvaged from Berkeley House in London, which had burned down in 1733. The present Monks' Garden stands on the site of the garden originally created in the twelfth century by the Bishops of Winchester; slightly further away from the house is the recently restored Temple of Diana.

The most celebrated occupants of Highclere Castle were the 5th Earl of Carnarvon, George Edward Stanhope Molyneux Herbert (1866–1923), and his wife Almina (1876–1969) who, during the First World War, transformed Highclere into a hospital for wounded soldiers. It was the 5th Earl who, along with his colleague, Howard Carter, discovered the tomb of Tutankhamun 'The Boy King' in the Valley of the Kings, Egypt, in 1922. Shortly after the discovery, the Earl died while in Cairo, and many claimed this to be the result of the Curse of Tutankhamun. The 5th Earl is buried within the ramparts of the former Iron Age hill fort on Beacon Hill – a fitting place for the adventurer, explorer and archaeologist; from the summit there is a wonderful panoramic view which includes Highclere Castle.

The castle, which is open during the summer, is full of antiques, including Napoleon's desk and chair. The State Dining Room is dominated by a painting of Charles I by Van Dyck, flanked on either side by portraits of Carnarvon ancestors who took part in the English Civil War. The Gothic-styled Salon has leather wall coverings from Spain and the magnificent library has close to 6,000 books, some dating back to the sixteenth century. In the cellar is the Egyptian Exhibition, opened to celebrate the 5th Earl's achievements in Egypt in finding Tutankhamun's tomb. And finally, just in case you think you've seen the impressive building somewhere before, Highclere Castle was the setting for the highly successful ITV drama *Downton Abbey*.

▲ Historic Littlecote House to the west of Hungerford

LITTLECOTE HOUSE

Situated to the west of Hungerford close to the River Kennet and within sight of the former Roman villa is Littlecote House (SU305703). The original house was built in the late thirteenth century by the de Calstone family, passing to the Darrell family on the marriage of Elizabeth de Calstone and William Darrell in 1415. It was one of William's descendants who built the present Tudor mansion in the early sixteenth century and it was here that Henry VIII met Jane Seymour, who became his third wife, shortly after Anne Boleyn was beheaded in 1536. The last Darrell at Littlecote was William, often referred to as 'Wild Darrell' as he was connected with several alleged scandals.

Legend has it that a local midwife was woken late one night in 1575 and offered a large sum of money by a masked gentleman if she would attend to a woman who was in urgent need of her services on the understanding that she had to be blindfolded. Sadly, following the birth, the 'unidentified gentleman' killed the baby by throwing it on the fire. The midwife kept her silence until, on her deathbed, she revealed the terrible secret. Suspicion fell on 'Wild Darrell'; however, there was never enough evidence to bring anyone to trial. Unfortunately for Darrell he was killed in a riding accident in 1589 at a spot that became known as Darrell's Stile.

Following the death of the last Darrell, Sir John Popham, the man responsible for sentencing Mary Queen of Scots to death, acquired Littlecote in 1589. He was later appointed Lord Chief Justice of England in 1592, a post he held for fifteen years. During this time he presided over the trials of Sir Walter Raleigh in 1603 and Guy Fawkes – the man behind the Gunpowder Plot – in 1606. Sir John was also responsible for the completion of the late-Elizabethan south façade of the present mansion, including the Great Hall; the rear of the house is the original Tudor structure built around 1500 by Sir George Darrell.

Within the Great Hall, with its Jacobean screen and ornate plaster ceiling, is a small stained-glass window containing a love-knot with the initials of Henry VIII and Jane Seymour; the couple are reputed to have courted at Littlecote. Jane Seymour was descended from Sir George Darrell – her father Sir John Seymour was the son of John Seymour and Elizabeth Darrell, daughter of Sir George Darrell.

During the Civil War Sir John's grandson Alexander Popham (who by then held the manor of Littlecote) favoured the Parliamentarians; however, he later changed his allegiance and helped in the restoration of Charles II.

The Popham and Leybourn-Popham families continued to own Littlecote for over 300 years until 1929, when Sir Ernest Salter Wills bought the house; the Wills family from Bristol made their fortune in tobacco, having founded the W. D. & H. O. Wills Tobacco Company, which became part of Imperial Tobacco in 1901.

Following the brief period during the Second World War when US soldiers, who were later involved in Operation 'Market Garden', were stationed at the house, the Wills family continued to live at Littlecote until the house was sold to Peter de Savary in 1985. The next chapter in Littlecote's history came in 1996 when Warner Holidays acquired the Grade I listed house; the house and its surrounding estate are now used as a large country house hotel and leisure resort.

TOTTENHAM HOUSE AND SAVERNAKE FOREST

Lying between Marlborough and Hungerford is Savernake Forest, an area of ancient woodland that pre-dates the Norman Conquest; it was mentioned as '*Safernoc*' in a Saxon Charter from King Athelstan in AD 934. After 1066 the hereditary wardenship of the forest was given to Richard Esturmy, a Norman knight, and Savernake Forest has been passed on in an unbroken line ever since. Although privately owned, it was leased to the Forestry Commission in 1939 on a 999-year lease and is open to walkers who want to explore the forest which contains many ancient trees, including the Big Belly Oak. This pollarded oak tree with a girth of 36 feet is said to be over 1,000 years old, making it one of the oldest trees in the forest.

In 1427 the wardenship of the forest was inherited by the Seymour family, one of whose descendants was Jane Seymour; in 1536 she became the third wife of Henry VIII and mother of his son and heir, Edward VI. In 1548 Edward Seymour, 1st Duke of Somerset (d.1552), was granted the full ownership of the estate following the death of Catherine Parr (sixth and last wife of Henry VIII) and sometime around 1570 his son took the decision to construct a grand manor house. Unfortunately, slightly over a hundred years later, the house was damaged by fire and the estate then passed by marriage to Lord Thomas Bruce (1656–1741) when he married Elizabeth Seymour in 1676.

Charles Bruce, 3rd Earl of Ailesbury (1682–1747), was next in line to hold the estate and he commissioned the architect Lord Burlington to build a new brick mansion in the early eighteenth century. Charles was succeeded in 1747 by his nephew, Thomas Brudenell (1729–1814), later Brudenell-Bruce, and the estate remained in the Brudenell-Bruce family until most of it was sold around 1930.

Thomas, Governor to the Prince of Wales (later George IV) and Prince Frederick, employed Lancelot 'Capability' Brown to redesign the surrounding parkland, which included long beech-lined avenues such as Grand Avenue through the heart of the forest in the 1760s; another addition was the Ailesbury Column in the 1780s (see later in this chapter). His son, Charles Brudenell-Bruce (1773–1856), succeeded him in 1814 and in 1821 was created 1st Marquess of Ailesbury. Charles commissioned the architect Thomas Cundy to redesign the mansion into its current grand Palladian form, built around the older Burlington-designed house using Bath stone (SU249639).

However, during the twentieth century, the grandeur of Tottenham House started to fade. During the Second World War, it was used by American Armed Forces based in Savernake Forest in preparation for D-Day, and then became home to Hawtreys Preparatory School until the 1990s. The house is currently awaiting redevelopment as an international golf resort.

WELFORD PARK

The present house and church at Welford Park (SU409731) stand on the site of a former monastery that was in the care of Abingdon Abbey until the Dissolution of the Monasteries. Henry VIII then kept Welford as a deer hunting lodge until he granted it to Sir Thomas Parry in 1546, who later became Treasurer of the Household of Elizabeth I. In 1618 the house was sold to Sir Francis Jones, one-time Lord Mayor of London. The present house was built for Sir Francis's grandson, Richard Jones, around 1652 by John Jackson of Oxford. Richard had no sons and his heiress, Mary, married John Archer in 1680; since then the house has passed by marriage and inheritance through the Archer, Eyre, Houblon and Puxley families, coming to the present owner, James Puxley, in 1997.

The main feature at Welford is the magnificent display of snowdrops in early spring, which is the only time people can visit. The snowdrops were probably cultivated by the monks at the original monastery for their religious and medical qualities; throughout early spring there are usually extensive displays of snowdrops in many ancient churchyards.

Snowdrop is the common name for members of the genus *Galanthus*, which comes from the Greek '*gala*' and '*anthos*' meaning milk-flower. During the Middle Ages snowdrops had strong religious associations, especially with the Virgin Mary, and were particularly associated with Candlemas; one of the old names for snowdrops was 'Candlemas bells'. The name Candlemas stems from the tradition of blessing all of the Church's candles for the coming year. This relates back to pre-Christian times when it was the festival of light, an ancient festival marking the mid-point of winter, halfway between the winter solstice (shortest day) and the spring equinox. There are many traditions and superstitions surrounding Candlemas, including one to predict the weather for the rest of the winter:

If Candlemas Day be fair and bright,
Winter will have another fight.
If Candlemas Day brings cloud and rain,
Winter won't come again.

The adjacent St Gregory's Church was almost completely rebuilt in the 1850s, with the exception of the rare Norman tower, one of only two round towers in Berkshire – the other being at St Mary's Church in Great Shefford. Inside there are several interesting monuments, the finest of which is next to the south door. This commemorates Lady Anne Parry (d.1585) of Welford Park; she is kneeling in prayer with her seven sons and twelve daughters below. There is also a beautiful arch-decorated late Norman or Early English circular font.

Interesting Churches

Churches form a quintessential part of the English countryside – who is not moved by the distant ring of church bells on a peaceful summer's day – and they offer a tantalizing glimpse into history through their architecture and memorials. Throughout the North Wessex Downs there are church towers and carvings that date back to Saxon times, doorways and fonts with beautiful Norman ornamentation, carved effigies, vivid stained-glass windows and some truly stunning medieval wall paintings, all waiting to be admired.

▲ Part of a Saxon carved grave cover in the Church of the Holy Cross, Ramsbury (see page 89)

Effigy of Sir Nicholas de la Beche – one of the 'Aldworth Giants' – inside St Mary's Church, Aldworth

Most churches should be open, or at least display contact details for a local key-holder, and there are usually printed booklets to buy that give much more detail than that provided here.

The churches listed here are not exhaustive, but cover some of the best features available; some other churches such as those at Ashmansworth, Great Bedwyn, Hungerford, Lambourn, Marlborough and Pewsey were covered in Chapter 4.

ALDWORTH – ST MARY'S CHURCH

The picturesque village of Aldworth, known as *Elleorde*, the 'Old Town', at the time of the Domesday Book, is home to the Bell Inn, housed in a building dating from around the thirteenth century, which has been owned by the same family for over 250 years. Opposite is a canopied well which, at 365 feet, is said to be the deepest in England. However, it is St Mary's Church (SU554793) that places Aldworth firmly on the history trail. The church, which dates back to Norman times, is famed for the 'Aldworth Giants', nine larger than life effigies of the de la Beche family dating from the fourteenth century. The family came to England in the wake of William the Conqueror and built a castle or fortified manor nearby, though today there is no trace of it. However, the De La Beche farm, to the south of Four Points, is believed to be built on the site of the castle. During excavations there in 1871 a silver seal bearing the name Isabella de la Beche was found; the seal is now held in Reading Museum.

The first of the stone figures along the north wall is that of Sir Robert de la Beche, who was knighted by Edward I in 1278. Then there is his son Sir John, and his son Sir Philip. Sir Philip, the largest figure in the church, was valet to Edward II before being appointed Sheriff of Berkshire and Wiltshire in 1314. However, in 1322, he took part in a rebellion led by the Earl of Lancaster to overthrow the King. The uprising came to an end at the Battle of Boroughbridge in Yorkshire; the Earl of Lancaster was beheaded and Sir Philip imprisoned. Five years later Isabella of France, once Queen to Edward II, returned to England with her son, who was heir to the throne. In 1327 he became Edward III and restored Sir Philip to the manor at Aldworth. The stone effigy of Lord Nicholas, Constable of the Tower of London and custodian to the first son of Edward III, the Black Prince, is in the nave, along with Sir John and his wife Isabella. The male line of the de la Beche family came to an end with the death of Lord Nicholas in 1348, though the family held the manor for another 150 years in the female line of Langford. Lord Nicholas' elder brother, Sir Philip, lies with his sister Lady Joan along the south aisle. The last figure, which has suffered the worst damage, is that of John, son of Lady Isabella, who died at the young age of 20 in 1340.

The damage to the effigies was probably done in the 1650s when Cromwell's Parliamentarian soldiers visited the church; however the overall state of preservation is impressive considering that they have been here for over 700 years. By 1640 the local people had given names to four of the stone figures; the largest was known as John Long, and the three others as John Strong, John Never Afraid and John Ever Afraid. The last, which has disappeared, is said to have promised his soul to the Devil, whether he was buried inside or outside the church. However, at his death he tricked the Devil, by being buried in the wall, neither in nor out of the church. The arch where the statue lay is on the outside of the south wall.

The area has connections with two poets: Alfred Lord Tennyson and Laurence Binyon. In 1850 Tennyson (1809–1892) married Emily Sellwood at Shiplake; Emily's family home was Pibworth Manor and her parents are buried in the churchyard (for information on Laurence Binyon see the Writers of Prose and Poetry section, page 150).

ALTON BARNES – CHURCH OF ST MARY THE VIRGIN

The name Alton, used by both Alton Barnes and its near neighbour Alton Priors, is derived from the Old English *aewielle-tun*, meaning farmstead by the stream or well, and a tributary of the Hampshire Avon rises between the two hamlets; the Barnes prefix is probably derived from the Berners family who held the manor after the Norman Conquest, while Priors received its prefix following the time when it was owned by St Swithun's Priory in Winchester.

At Alton Barnes, overlooked by the white horse high on the downs, is the Church of St Mary the Virgin (SU107620) which dates back to Saxon times, although it was mostly rebuilt in the eighteenth century. Inside are a fine sixteenth-century tie-beamed and wind-braced roof, a Georgian gallery and some interesting monuments.

ALTON PRIORS – ALL SAINTS CHURCH

Just a short walk across the field from the church at Alton Barnes is the redundant All Saints Church in

▲ One of two sarsen stones beneath the floor of All Saints Church in Alton Priors

Alton Priors (SU109621), now in the care of the Churches Conservation Trust. The church dates from Norman times, although all that remains from that period is the chancel arch; the rest is late medieval with the exception of the chancel which was rebuilt in the nineteenth century. Inside there are some fine Jacobean carved wooden choir stalls and a tomb chest to William Button (d.1590). However, the reason many come to visit is to see the two large sarsen stones buried beneath the floor that are visible through two wooden trapdoors. Maybe these are from an earlier sacred site – early Christian churches were sometimes built on existing religious sites. Outside in the churchyard stands an ancient yew tree, reputed to be 1,700 years old, another pagan symbol that is many centuries older than the church.

ASHAMPSTEAD – ST CLEMENT'S CHURCH

The village of Ashampstead, which dates back to at least the time of the Norman Conquest, when William the Conqueror gave the lands to William FitzOsborn, is home to St Clement's Church (SU564767). The building dates from the late twelfth century, with the tower being added in the fifteenth century; the impressive timber roof and the weather-boarded bell-turret date from the sixteenth century. However, the real treasure of this church is to be found inside, where the walls are decorated with some of the finest thirteenth-century medieval wall paintings to be found in Berkshire.

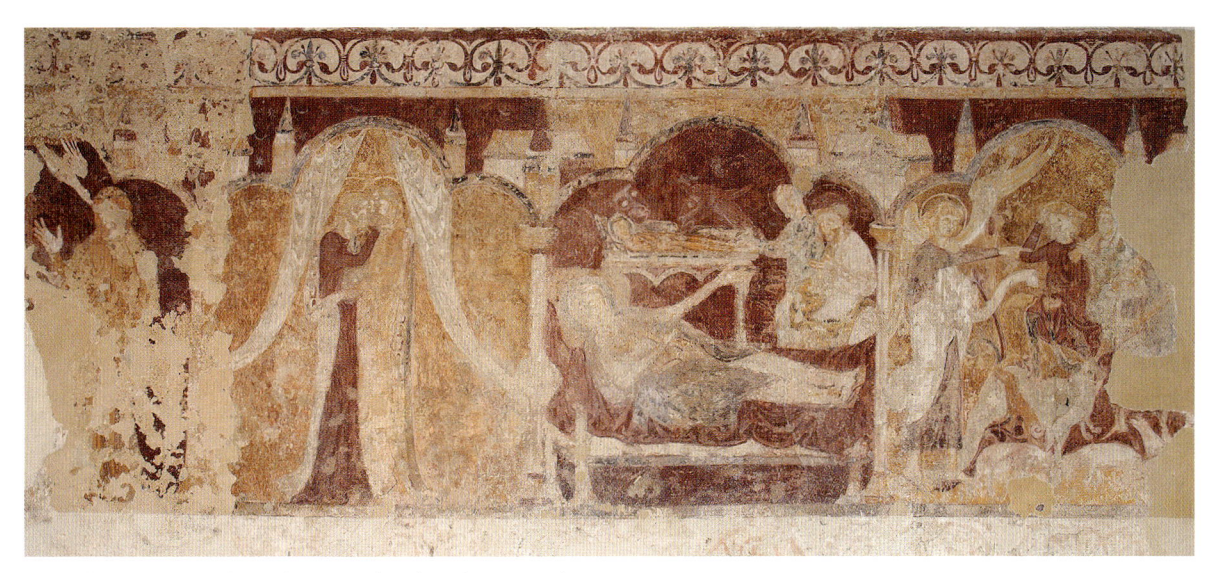

▲ Wall paintings inside St Clement's Church, Ashampstead

The painting on the north wall of the nave depicts religious scenes from the birth of Christ: the Annunciation; the Visitation; the Nativity with Jesus in the manger; and the Appearance of the Angel to the shepherds (going from left to right). On the chancel arch (unusually, it's not an arch as such, but a large timber beam) can be seen the remains of the Last Judgement, or Doom, with the figures of Christ and the Apostles over the chancel arch and souls being admitted to Heaven (left) or dragged down to Hell (right). The paintings were plastered over after the Reformation (sixteenth century) and were only rediscovered in 1895.

AVEBURY – ST JAMES' CHURCH

The actual village of Avebury is often overlooked by visitors who come to admire the ancient relics that our Neolithic ancestors have left us. However, to do this is to miss out on some rather picturesque thatched and tiled cottages as well as the manor house (see page 116) and the interesting St James' Church (SU099699). Originally dedicated to All Saints in the thirteenth century, the church is over a thousand years old and still retains its tall Anglo-Saxon nave, albeit altered by the Normans. The Norman aisles were added to the Saxon structure in the twelfth century and the tower is fifteenth century.

Inside, the font is thought to be of Saxon origin, with later Norman carving displaying intersecting arches with – depending on which interpretation you believe – either Christ trampling on two dragons, or serpents trying to bite the cloak of the bishop holding a crosier. On the south wall is a memorial to John Trusloe (d.1593), who held the manor of nearby Avebury Trusloe from 1568, while in the tower is a Royal Arms of George III, used before 1801, and a thirteenth- or fourteenth-century stone coffin thought to be that of a prior from the adjacent Benedictine priory.

One of the main features of the church is the rather grand and quite rare fifteenth-century rood loft; this was originally used to store the rood, or large crucifix, the most revered object in the early Church. Following the sixteenth-century Reformation the screen and lofts of many churches were removed and the timber reused elsewhere. However, in Avebury's case, the timbers were concealed within the church, only being rediscovered in 1810. The rood loft was restored during late nineteenth-century renovations and a new lower panel with paintings of the Apostles was added.

AVINGTON – CHURCH OF ST MARK AND ST LUKE

Just to the north-west of Kintbury, across the River Kennet and the Kennet and Avon Canal, is the little hamlet of Avington. Here (SU372679) is a small church – the Church of St Mark and St Luke – that dates entirely from the twelfth century with the exception of a sixteenth-century porch and nineteenth-century vestry.

All but one of the windows are Norman, following a simple round-headed style, while the south doorway is a richly carved affair with Norman zigzag decoration and carved capitals. Just inside is the barrel-shaped font carved with an arcade of eleven arches and containing religious figures. Behind the font is the blocked-up north doorway, which contains a grave slab incised with a cross; the north doorway, or 'Devil's door', was built to let out evil spirits during baptisms. However, it is the richly carved chancel arch, which over the centuries has sagged, that draws the eye. This has carved capitals and two series of decorative motifs including zigzags, scallops and beak heads (aptly named for they look like heads with a beak).

BISHOPS CANNINGS – PARISH CHURCH OF ST MARY THE VIRGIN

Right on the western edge of the North Wessex Downs, tucked into the Vale of Pewsey, stands the tall spire of St Mary's Church (SU037642), clearly visible from miles around. This large church, which dates from around 1150 – with considerable rebuilding in the fifteenth century, at which time the impressive spire was added to the tower – appears to be out of keeping with the small village of Bishops Cannings; however, this was always an important parish, linked to the Bishopric of Salisbury. After admiring the spire, go inside to see the finely carved pew ends from the 1880s – each one is different – and take a look at the rare seventeenth-century penitent's pew. This is where local parishioners would sit beneath the large painted hand – the 'hand of God' or 'hand of meditation', with warnings about vanity, mortality and sin – while they pondered their sins.

▲ Parish Church of St Mary's, Bishops Cannings

FROM STATELY HOMES TO WHITE HORSES

▲ The carved pew ends inside St Mary's Church date from the 1880s

▲ The rare seventeenth-century penitent's pew inside St Mary's Church

BOXFORD – ST ANDREW'S CHURCH

The picturesque village of Boxford, through which the River Lambourn flows, is home to St Andrew's Church (SU428716). Parts of the underlying structure of the church date back to Saxon times, though most of what can be seen is from the early thirteenth century. The tower was added in the fifteenth century; however, from the records of Oliver Sasson, a Quaker who lived in an adjacent house, it is known that the original tower collapsed in the seventeenth century and was replaced some years later by the current brick tower. The interior of the nave and chancel were extensively renovated around 1910. Yet, St Andrew's Church has a 'hidden' gem, only rediscovered during repair work undertaken in 2010.

While concrete render was being removed from the exterior of the flint walls of the nave and chancel, a previously unknown window in the south wall of the chancel was uncovered. Further investigations revealed a small timber window frame with hinged wooden panel built by Saxon craftsmen more than a thousand years ago. The discovery has allowed the picturesque rural church to lay claim to having the oldest working wooden window in England.

EAST SHEFFORD – CHURCH OF ST THOMAS

Little remains of the village of East Shefford; the original manor house was demolished in 1871, the present building being a more recent addition. However, close to the banks of the River Lambourn is the little Church of St Thomas (SU390746), which is now cared for by the Churches Conservation Trust. There is not much to see on the outside of the Norman church, which dates from around 1100; the south chapel was added in

▲ Monument to Sir Thomas and Lady Beatrice Fettiplace in the Church of St Thomas, East Shefford

▲ Inside the Church of St Thomas there are several faded wall paintings, including the scene of the Nativity on the chancel arch

▲ The engraved window by Sir Laurence Whistler in memory of the poet Edward Thomas inside the Church of St James the Greater in Eastbury

the fifteenth century. However, the development of the church, and village, reflects the fortunes of the Fettiplace family who held the manor house. The Fettiplaces were first mentioned in 1223 when Adam Fettiplace held the Manor of Oxford. The first family member to live at East Shefford was Sir Thomas Fettiplace (d.1442) who married Lady Beatrice (d.1447) from the Royal House of Portugal. The church stands as a private mausoleum to the Fettiplaces and includes several interesting monuments. The one to Sir Thomas and Lady Beatrice Fettiplace (c.1450) in the south chapel is particularly good; Sir Thomas' armour is typical of the reign of Henry VI. Along the north wall is a canopied tomb of John and Dorothy Fettiplace (c.1524); their kneeling effigies are in brass. The church is decorated with early wall paintings and texts – though these are very faded, they include a scene of the Nativity on the chancel arch.

EASTBURY – CHURCH OF ST JAMES THE GREATER

The picturesque village of Eastbury, with its collection of sixteenth- and seventeenth-century thatched cottages, was first mentioned in 1164 when it was known as *Estberi* and the estate was held by Ralph de Lanvalei. Although the Church of St James the Greater (SU345771) may look old, it was really built in the mid-1850s, replacing an earlier church that had fallen into disrepair. The church, along with the school and master's house, were all commissioned by Robert Milman, Vicar of Lambourn in the mid-nineteenth century; the architect was George Edmund Street.

The most interesting feature is the engraved window by Sir Laurence Whistler (1912–2000), a fitting tribute to the poet Edward Thomas (1878–1917) and his wife, Helen, who originally lived in the village of Steep in Hampshire. After Thomas was killed in action at the Battle of Arras, on the western front, on Easter Monday 1917, his widow moved to Eastbury, where she remained until her death in 1967. On the left of the window is a tree in leaf with the initials and dates of both Edward and his wife carved into the bark, while to the right is a winter tree, devoid of leaves, where Edward's First World War helmet and belt hang from the branches. Interspersed are lines of poetry, views of the rolling chalk downs, the spire of Steep Church and, in the far distance, the rugged Welsh mountains from his homeland – the more you stand and admire, the more you see.

▲ All Saints Church, Farnborough

FARNBOROUGH – ALL SAINTS CHURCH

The little village of Farnborough, which hides away high in the Berkshire Downs at 735 feet, was once the home of the poet laureate Sir John Betjeman (1906–1984). Betjeman, often described as the 'twentieth-century's most popular poet', loved walking on the downs and moved to the village with his wife, Penelope, in 1945. They lived at the Old Rectory, a picture-postcard house built in the eighteenth century, before moving to Wantage in 1951. Close by is All Saints Church (SU435819), essentially twelfth-century, though the tower was added in the fifteenth century and the church underwent major restorations in the late 1880s. The colourful, stained-glass west window, depicting symbols of the Resurrection – a tree of life flanked by fish and butterflies – was designed by the painter John Piper (1903–1992) in memory of his friend Sir John Betjeman. The inscription mentions that the window was made by Joseph Nuttgens and placed there by the 'Friends of Friendless Churches'.

▲ Colourful stained-glass window dedicated to Sir John Betjeman in All Saints Church

▲ Twelfth-century coffin lid inside St Mary's Church, Hampstead Norreys

▲ Hannington's village green with the well head built in 1897 to celebrate Queen Victoria's Diamond Jubilee

HAMPSTEAD NORREYS – ST MARY'S CHURCH

At the time of the Domesday Book the village of Hampstead Norreys was known as *Hanstede* – meaning 'farm settlement', and was owned by 'Theodric the Goldsmith'. In 1450 the parish was renamed Hampstead Norreys (sometimes spelt 'Norris') when the manor and village lands were bought by the Norreys family from Bray. The Church of St Mary the Virgin (SU529762), which is Norman with Early English and Perpendicular additions including the fifteenth-century tower, was built on the site of an earlier Saxon structure. Inside, the chancel has a solid fifteenth-century wagon roof, while on the south wall is a stone slab which was found during restorations in 1880. The slab, probably a coffin lid, is carved with the figure of a knight on horseback and, judging by the shape of the cylindrical helmet, it could date from the late twelfth century. There are also the remains of a few faded wall paintings, probably from the thirteenth century.

HANNINGTON – ALL SAINTS CHURCH

The beautiful little village of Hannington, whose name derives from *Haningtun*, meaning the 'farm of Hana', hides 660 feet up in the North Hampshire Downs. Clustered around the large village green, with its pyramidal-roofed well head that was built in 1897 to celebrate Queen Victoria's Diamond Jubilee, are a lovely collection of brick and timber cottages. Beyond the lychgate lies All Saints Church (SU538554) which dates back to Saxon times.

If you get a chance to go inside, have a look at the two beautifully engraved memorial windows by Sir Laurence Whistler – a leading exponent of hand engraving during the twentieth century. One window remembers William Whistler (1886–1978), whose family have farmed at Hannington for generations. The window shows the Scythe of Time, with William's dates of birth and death on the two handles; a sheaf of corn and flock of sheep being driven by a border collie

FROM STATELY HOMES TO WHITE HORSES

▲ Engraved window by Sir Laurence Whistler in memory of a local farmer, William Whistler (no relation), in All Saints Church, Hannington

▲ Engraved window detail by Sir Laurence Whistler in memory of Rose Hodson in All Saints Church, Hannington

illustrate his life as a farmer on the downs. The other window remembers Rose Hodson (1926–1986) and depicts the house, built in 1793, where she lived for twenty-two years.

LOWER BASILDON – ST BARTHOLOMEW'S CHURCH

The redundant Church of St Bartholomew (SU611792), now owned and cared for by the Churches Conservation Trust, was, at the time of the Domesday Survey of 1086, part of a Royal Manor. It was later acquired by the abbey at Lyre in Normandy and remained under their control until 1337, when Edward III abolished such foreign rights at the beginning of the Hundred Years' War with France. It was during the 200 years of French ownership that much of the present church was built. The nave, which dates from 1220, is constructed in an Early English style, whereas the chancel, which was rebuilt about sixty years later, is a good example of the Early Decorated style. The present brick tower, typical of some churches in Berkshire, was built in 1734. Inside there are several interesting memorials, including a fifteenth-century floor brass, with two figures in medieval costume, commemorating John Clerk (d.1497) and his wife Lucy. However, most of the memorials are to previous owners of the manor of Basildon, including Roger Yonge (d.1589), a Bristol merchant who bought the manor in 1543. There is another to Sir Francis Sykes (1730–1804) from Yorkshire, who made his fortune after joining the East India Company; Sykes bought the Basildon estate in 1768 and built the present Basildon House.

▲ The Church of St Bartholomew at Lower Basildon

In the churchyard is a memorial to the famous agricultural innovator Jethro Tull (see From the Land to Science – page 60). The memorial stone was placed against the south wall of the church by Gilbert Beale, the founder of Beale Park, in 1941, 200 years after Tull's death. Close to the main door is a touching memorial to the two sons of farmer Edward Deverell from the adjacent Church Farm, who drowned in the River Thames in 1886.

MILDENHALL – ST JOHN THE BAPTIST CHURCH

Mildenhall, pronounced and sometimes written as 'Minal', lies in the Kennet Valley just to the east of Marlborough, close to the site of the former Roman town of *Cunetio*, and is home to the Church of St John the Baptist (SU209695).

The church, which was mentioned in the Domesday Book, dates back to Saxon times; look closely at the lower parts of the tower and there are traces of Saxon stonework and a blocked-up window. While outside, take a look at the sundial which probably dates from the early nineteenth-century restorations; sundials were originally designed to show the local time as calculated by their latitude, so the time here is 7 minutes behind Greenwich Mean Time (GMT).

Inside, the six main arches and corresponding large columns in the nave date from the twelfth century, while in the east window are fragments of stained glass claimed to be some of the oldest in Wiltshire. Above the chancel arch are the early nineteenth-century Royal Arms of George III, used to symbolize the tie between Church and State.

However, it is the magnificent Georgian woodwork, fortunately left untouched by Victorian restorers, that places Mildenhall's church high on a pedestal. In 1816, twelve wealthy parish members funded the refurbishment of the interior that we see today, with beautiful shoulder-high box pews, matching pulpit and reading desk and a semi-circular gallery at the west end.

It was Sir John Betjeman who referred to St John's as 'a church of a Jane Austen novel' and having 'a forest of magnificent oak joinery', while the artist John Piper made reference to the colour of the wood as that 'of an old fiddle'.

STANFORD DINGLEY – CHURCH OF ST DENYS

The name of the picturesque village of Stanford Dingley is derived from the original lord of the manor William de Stanford, mentioned in 1224, and the Dyneley family who lived here in the Middle Ages; there is a brass memorial to Margaret Dyneley dated 1444, in the church. The church (SU575716) is unusual in that it is dedicated to St Denys, who was martyred in third-century France; legend has it that after being beheaded his body stood up, picked up his head and walked through the streets to his chosen burial spot. The church has Saxon origins, though it mostly dates from the twelfth and thirteenth centuries; the white weatherboarded bell turret being added in the fifteenth century. During the thirteenth century the church interior was decorated with wall paintings and some fragments of these can still be seen. A more recent addition is the engraved memorial window by Madeleine Dinkel to the novelist and poet Robert Gathorne-Hardy (1902–1973), who lived at Mill House in the village for many years. Another well-to-do village resident was Thomas Tesdale (1547–1610) who, from his personal fortune, co-founded Pembroke College in Oxford. There is also a memorial brass to John Lyford (d.1610) who lived in the parish that states he was a citizen and merchant tailor of London.

▲ Former school buildings at Upper Woolhampton (now private houses)

UPPER WOOLHAMPTON – DOUAI ABBEY

Following Henry VIII's Dissolution of the Monasteries many people who had led a monastic lifestyle fled abroad and, in 1615, one group of monks founded the community of St Edmund in Paris. In the wake of the French Revolution the survivors of St Edmund's moved to the town of Douai in Flanders (now in northern France) in 1818. However, owing to the changing political situation in France at the end of the nineteenth century, the Benedictine community of St Edmund were expelled from Douai in 1903. The Bishop of Portsmouth offered them a site at Upper Woolhampton and it was here that they combined their school, St Edmund's College, with the existing St Mary's College to form Douai School, which became a leading Roman Catholic public school. The school finally closed in 1999 and most of the buildings, which had been constructed during the nineteenth century in a Gothic style, were converted into flats and houses. However, the adjacent abbey church (SU577681) is still home to the Douai community of monks. Work on the abbey church,

▲ The distinctive white weatherboarded turret of St Denys' Church, Stanford Dingley

▲ The unusual Douai Abbey at Upper Woolhampton

▲ The interior of Douai Abbey at Upper Woolhampton

▲ The rare pre-Conquest tower at Swithun's Church, Wickham

WICKHAM – ST SWITHUN'S CHURCH

Wickham's St Swithun's Church (SU394715) has one of the best surviving Anglo-Saxon towers in Berkshire. The hilltop location and the design of the sturdy pre-Conquest tower – the entrance at first floor level would have been reached by a ladder which could be drawn up in times of trouble – indicates that the tower may have originally been a defensive structure. The rest of the church is Victorian, having been almost entirely rebuilt in the period 1845–9 by the rector, William Nicholson. Inside there is a fascinating group of eight papier-mâché elephants decorating the beams of the roof in the north aisle. Nicholson bought three elephants at the Paris Exhibition in 1862, with the original intention of putting them in the rectory (now Wickham House); however, unable to find a suitable location, he had five more made and then placed them in the church. Originally gilded, the elephants still make a striking contribution to the interior of the church.

The church is named after St Swithun: born around

designed by Arnold Crush in a Gothic revival style, began in 1929. The construction was stopped in 1933, with only the eastern end of the present structure having been built. Work started on completing the abbey church in 1987, following a much more modern style; the result is a striking mixture of styles with a bright interior bathed in light.

Monuments, White Horses and Crop Circles

Throughout the North Wessex Downs people have placed memorials, monuments and white horses, including one that's around 3,000 years old, as both memorials and bold statements; military memorials are covered in Chapter 2. The region is also well known for a proliferation of crop circles.

MONUMENTS AND MEMORIALS

Ailesbury Column

Situated on an area of higher ground in the southern half of Savernake Forest is the Ailesbury Column (SU229648). The column, topped with a bronze urn, was erected in 1781 by Thomas Bruce, Earl of Ailesbury, although it is believed to have been originally set up at Brandenburg House, Hammersmith, in the 1760s. One of the two plaques on the monument commemorates the restoration 'to perfect health from a long and afflicting disorder' of King George III in 1789.

▲ One of the eight papier-mâché elephants that adorn the roof inside St Swithun's Church at Wickham

the start of the ninth century he later became chaplain to Egbert, the first king to unite the kingdoms of England. In AD 852 he was made Bishop of Winchester by Egbert's son Ethelwulf. Many will have heard of St Swithun from the folklore tradition about St Swithun's Day (15 July):

St Swithun's Day, if thou dost rain,
For forty day it will remain:
St Swithun's Day, if thou be fair,
For forty days 'twill rain na mair.

On his death, in AD 862, St Swithun left instructions that he wished to be buried in a common graveyard without any ceremony. However a century later his remains were taken to Winchester Cathedral. On the day his remains were to be removed, 15 July, it rained so hard that the ceremony had to be postponed. It continued to rain for forty days and this was taken as a sign of the saint's displeasure at having his wishes disobeyed.

▲ The Ailesbury Column in Savernake Forest

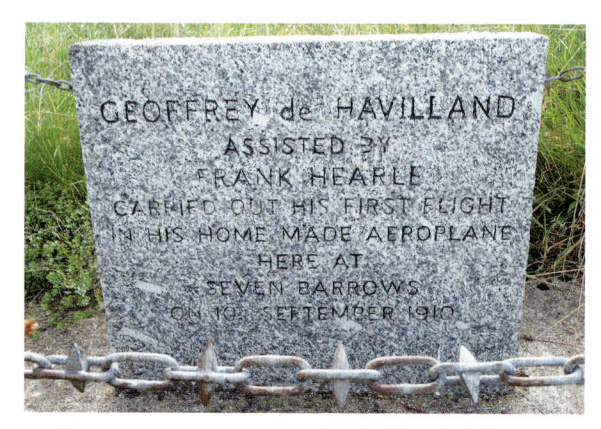

▲ Memorial stone commemorating the first flight by Sir Geoffrey de Havilland

▲ Combe Gibbet stands high up on the North Hampshire Downs

Beacon Hill and Seven Barrows

We have already discovered that the grave of the 5th Earl of Carnarvon (Highclere Castle) is sited within the former Iron Age earthworks on Beacon Hill (SU458572). However, 1¼ miles to the south, beside a group of Bronze Age barrows known as Seven Barrows and the busy A34, is a small memorial stone. This commemorates the first successful flight on 10 September 1910 by Sir Geoffrey de Havilland (1882–1965), pioneer aviator, aeroplane designer, and founder of the aircraft company that bore his name (SU461550).

Burderop Down

On Burderop Down (SU158762), just a short stroll from Barbury Castle, is a large sarsen stone with memorial plaques erected to the memory of two local writers: Richard Jefferies (1848–1887) and Alfred Williams (1877–1930), who were inspired by the beautiful landscape around them. Richard Jefferies, who lived at Coate, was a keen naturalist, while Alfred Williams, a poet, lived at South Marston, both near Swindon (for more information about these two writers see pages 149–50).

The plaque to Richard Jefferies reads: 'It is eternity now. I am in the midst of it. It is about me in the Sunshine', a quotation from his autobiography, *The Story of my Heart.*

That to Alfred Williams reads: 'Still to find and still to follow, joy in every hill and hollow. Company in solitude.'

Combe Gibbet

Standing as a grisly reminder of a bygone era, rather than a memorial in the usual sense, is the haunting outline of Combe Gibbet (SU364622), high up on Inkpen Beacon, clearly visible for miles around. George Broomham, a local farm labourer, and his mistress, Dorothy Newman, were tried at Winchester Assizes in 1676 for the murder of George's son and wife. Following the guilty verdict, the pair were executed at Winchester before being hanged in chains close to the scene of the murder as a deterrent to others.

▲ Walkers passing the stark outline of Combe Gibbet

Fortunately the double gibbet, sited on a Neolithic long barrow, was never again used to hang anyone, although it has been replaced on several occasions. From the summit ridge there are stunning views both north across the Kennet Valley and south over the Hampshire Downs.

The story of the murders was used as the basis of the 1948 film *Black Legend*, which was produced by a group of Oxford undergraduates, including John Schlesinger (1926–2003) who later became a well-known Oscar-winning film director: he regarded *Black Legend* as 'his first success'.

Lady Penelope Betjeman's Memorial

Just off the north side of the Ridgeway (through a gate), not far from the monument to Robert Lloyd-Lindsay, is a sarsen stone and plaque (SU442849) in memory of Lady Penelope Betjeman (1910–1986) – wife of the former poet laureate Sir John Betjeman – who enjoyed riding over the downs.

Lansdowne Monument

In the farthest west corner of the North Wessex Downs, near Cherhill in Wiltshire, is the Lansdowne Monument (SU047693), also known as the Cherhill Monument. The 125-foot stone obelisk was built by the 3rd Marquess of Lansdowne in 1845 to the designs of Sir Charles Barry to commemorate his ancestor, Sir William Petty (1623–1687), a well-known seventeenth-century economist, scientist and philosopher.

Unfortunately, as a result of its exposed location high on the downs, the monument has suffered damage and, at the time of writing, is surrounded by scaffolding. Although the National Trust has plans to repair the monument, no date has been given.

Robert Loyd-Lindsay's Memorial

Situated on the Ridgeway high up above Wantage (SU423843) is a marble column topped with a cross commemorating Colonel Robert Loyd-Lindsay (1832–1901), who was made Lord Wantage of Lockinge in 1885. Loyd-Lindsay was a famed soldier during the Crimean War, taking part in the Battles of Alma and Inkerman in 1854, and was awarded the Victoria Cross. He was a founding member of the British Red Cross, and was also responsible for the statue of Alfred the Great in Wantage, although he is probably better known locally for developing the Lockinge Estate.

Robert James Lindsay, as he was originally known, married Harriet Loyd, the only child of Lord and Lady

▲ The memorial to Lady Penelope Betjeman

▲ Lord Wantage Monument on the Ridgeway high above Wantage

Overstone, in 1858 and adopted the name Loyd-Lindsay. Lord Overstone gave both the Lockinge Estate and Lockinge House to the couple and they set about creating a new Victorian estate around Lockinge and Ardington; by 1873 the estate comprised around 20,000 acres, making it not only the largest estate in Berkshire but one of the largest in England. Lockinge House was demolished in 1947, though the village still shows off its Victorian estate architectural heritage. Lord Wantage is buried at Holy Trinity Church in Ardington.

Swanborough Tump

In the heart of the Vale of Pewsey, about 2 miles west of Pewsey, is the Swanborough Tump (SU130661); a 'tump' being a small mound or stand of trees. A plaque mentions that around AD 850 this place was known as *Swinbeorg* and was the meeting place of the Hundred of Swanborough. It was also here that, in AD 871, Alfred the Great and his elder brother, King Ethelred I, met up on their way to fight the invading Danes.

Wilder's Folly

Tucked away in a far eastern corner of the AONB beside Nunhide Farm, not far from the busy M4 motorway, is Wilder's Folly (SU647725), also sometimes known as Nunhide or Pigeon Tower. The simple circular brick folly was built in 1769 by the Revd Henry Wilder of Sulham House while he was courting Joan, younger sister of John Thoyts of Sulhamstead House.

He chose the location so that the tower could be seen from both Thoyts' home to the south across the River Kennet and from his own residence, Sulham House. The tower has an open ground-level arcade and the upper floors were accessed by an external wooden staircase; during the latter half of the nineteenth century the windows were bricked up so that the tower could be used as a dovecote.

WHITE HORSES

Probably the most visually striking 'monuments' to be found within the North Wessex Downs are the carved white horse hill figures of which there are eight: at Alton Barnes, Broad Town, Cherhill, Devizes, Hackpen, Marlborough, Pewsey and Uffington.

Alton Barnes

The Alton Barnes White Horse, which is 160 feet long and 166 feet high, is located on the Pewsey Downs to the north of the village of Alton Barnes, not far from Adam's Grave, a Neolithic long barrow, between Milk Hill and Tan Hill (SU106637). Robert Pile paid a travelling painter, John Thorne, £20 in 1812 to design the

▲ Wilder's Folly near Sulham

▲ Alton Barnes White Horse looks out over the Vale of Pewsey

horse and he employed John Harvey of Stanton St Bernard to cut it. Unfortunately for Robert Pile, Thorne fled before the work was finished and he was left to pay out again. However, John Thorne's misfortune was far worse; he was later arrested and charged with a series of crimes for which he was hanged.

In recent years the horse has been lit with candles to mark the winter solstice on 21 December, marking the return of the sun and the days getting longer. In the neighbouring village of Alton Priors there is a sarsen stone by the roadside onto which has been carved a miniature version of the Alton Barnes White Horse.

Broad Town

The fairly small Broad Town White Horse (80 feet long by 60 feet high), which overlooks the village (SU098783), was said by the Revd William C. Plenderleath, who wrote the book *White Horses of the West of England* (1885), to have been completed in 1864 by a local farmer William Simmonds of Littletown Farm. However, a curator from the Imperial War Museum claimed in a newspaper interview in 1919 that, as a schoolboy in 1863, he had helped scour the horse and that at that time it was at least fifty years old. As to which story is true is anybody's guess, although most believe Plenderleath's account is the more plausible: maybe the curator meant the nearby Hackpen Hill White Horse or was mistaken about the dates.

In 1991 the Broad Town White Horse Restoration Society was formed to undertake the restoration of the horse, which involved 200 tons of chalk being placed on the figure, and they continue to scour it regularly.

Cherhill

The Cherhill White Horse, situated on the edge of Cherhill Down just east of the village of Cherhill, close to the Iron Age earthwork of Oldbury Castle and the Lansdowne Monument (SU049695), is the second oldest of the Wiltshire horses. It was cut in 1780 under the direction of Dr Alsop of Calne, sometimes referred to as 'the mad doctor', who shouted instructions over a megaphone from the main road. Strangely enough, the eye was once filled with upturned bottles which used to sparkle in the sunlight, making it visible over a considerable distance (although by the time the Revd Plenderleath visited in the 1870s the bottles had gone); the horse, which is 123 feet long and 131 feet high, was fully restored in 2002 by the Cherhill White Horse Restoration Group.

▲ The Cherhill White Horse and Lansdowne Monument (see page 139)

Devizes Millennium White Horse

In 1999 Wiltshire gained its eighth and newest chalk horse figure which was created to celebrate the new Millennium. Designed by Peter Greed, the figure was cut by around two hundred local people on Roundway Hill to the north of Devizes, overlooking the village of Roundway (SU016641). The horse faces towards the Alton Barnes horse, making it the only chalk horse in Wiltshire, and one of only four in Britain, to face to the right.

However, this was not the first white horse in Devizes. The old horse, which is no longer visible, was cut in 1845 by the local shoemakers, and was known as 'Snobs Horse'; 'snob' being a local word for shoemaker. But towards the end of the nineteenth century the horse, which was located just below the Iron Age earthwork of Oliver's Castle (SU000645), had all but disappeared. There were several proposals to restore the figure, though none came to fruition and eventually it was decided that a new Millennium horse would be cut instead.

▲ Hackpen White Horse to the south-east of Broad Hinton

Hackpen

The Hackpen White Horse to the south-east of Broad Hinton on Hackpen Hill just below the Ridgeway (SU128748) was cut in 1838 by Henry Eatwell, parish clerk of Broad Hinton, to commemorate the Coronation of Queen Victoria. The horse measures 90 feet long by 90 feet high.

Marlborough

The Marlborough White Horse, sometimes known as the Preshute White Horse, lies just to the south of Marlborough College on Graham Hill (SU184682). The horse, which is the smallest in Wiltshire at just 62 feet long and 47 feet high, was designed by William Canning, son of Thomas Canning of the Manor House in Ogbourne St George, who was a pupil at Mr Greasley's Academy on the High Street in Marlborough (this later became the Ivy House Hotel and after closing has been converted to give extra boarding accommodation for Marlborough College), and was cut by a group of schoolboys in 1804.

The chalk figure was scoured every year until the school closed in 1830; despite being neglected on a few occasions, the horse still exists, although its shape appears to have changed over the intervening 200 years.

Pewsey

Pewsey, like Devizes, has had two white horses, although the original one has long since disappeared. The original horse was cut by Robert Pile of Manor Farm, Alton Barnes, probably around 1785 and records show that the horse was scoured in 1789. However, it appears that the landowner objected to the scouring festivities and refused to allow it again. The hill figure fell into neglect; by the mid-1800s it was in a very poor state of repair and by the twentieth century was no longer visible. Local tradition holds that the horse originally had a rider; however, there appears to have been no visible rider in the late 1800s when the chalk was still visible, so whether there was a rider or not, no one really knows.

Pewsey's second horse which, like most horses, faces left, was designed by George Marples and cut by the

Pewsey Fire Brigade to celebrate the Coronation of King George VI in 1937. The horse, which measures 66 feet long and 45 feet high, is on Pewsey Hill to the south of the town looking out across the Vale of Pewsey towards the Alton Barnes horse (SU171580).

Uffington

The Uffington White Horse, which lies just to the north of the Ridgeway along the northern scarp of the downs looking out across the Vale of White Horse (SU301866), is by far the oldest horse hill figure in the country. Although it was first mentioned in a medieval manuscript from Abingdon Abbey – *'juxta locum qui vulgo Mons Albi Equi nuncupatur'* ('near the place which is commonly called White Horse Hill') – recent excavations and new dating techniques have shown that the horse was carved some 3,000 years ago in the Bronze Age, somewhere between 1400 and 600 BC.

It was G. K. Chesterton (1874–1936), in his *Ballad of the White Horse*, who summed up the age of the horse perfectly:

Before the gods that made the gods,
Had seen their sunrise pass,
The White Horse of the White Horse Vale,
Was cut out of the grass.

The figure's remarkable state of preservation has been put down to the 'scouring fairs' that used to be held every seven years, when local people would gather to help clean and repair the figure, stopping the grass from growing over it. The first written record of the scouring was by Thomas Baskerville, in his Travel Journal of 1667. Thomas Hughes, who spent his childhood in the village of Uffington, wrote about the area in *Tom Brown's Schooldays* (1856), and also about the scouring rituals in *The Scouring of the White Horse* (1859); the small museum in Uffington is worth a visit for those interested in his life and work.

Following the First World War, it was reported that the horse was in a rather neglected state and during the Second World War the horse, like all other chalk hill figures, was covered over so that German bombers could not use it as a navigational aid. The horse was cleaned in the 1950s and the National Trust now undertakes the role of looking after this ancient landmark.

The small, flat-topped, mound of Dragon Hill far below the white horse is where St George is reputed to have killed the dragon. The bare patch of ground is said to have been caused by the dragon's blood, poisoning the soil forever. The steep-sided coombe, or dry valley, to the left of Dragon Hill, is known as The Manger, and legend has it that the white horse goes there to feed.

The highly stylized figure galloping across the downs is formed from deep trenches filled with crushed white chalk and measures 374 feet from end to end, making it not only the oldest but also the largest white horse around.

We may be able to determine the age of the horse with some degree of accuracy these days; however, as to its purpose, well that's a much harder question to answer. Maybe it had some religious significance, or was an emblem of a local tribe, cut to mark their land – similar representations have been found on coins from the Iron Age – the simplest answer is that most likely we'll never really know what its real purpose was.

And just in case you are wondering … the best view is from above.

Others that have disappeared

High up on the steep, north-facing scarp of the North Hampshire Downs was a white horse known as the Ham Hill or Inkpen horse (SU348621). The horse, which was shown on a late nineteenth-century Ordnance Survey map, is believed to have been cut by a former owner of Ham Spray House, yet today there is no trace of it.

Another long since disappeared figure was the Rockley Horse (SU152732), located high up on Rockley Down. Nothing is known about the origins of the horse; certainly the Revd William Plenderleath made no mention of it when he wrote about the other white horse hill figures in the area. The horse, which measured 126 feet from head to tail, was only 'discovered' when the field was ploughed in the 1940s, revealing the chalk on the surface of the soil; the figure is now completely lost.

CROP CIRCLES

More modern visual phenomena than white horses are crop circles – or maybe we should say crop formations, as they are certainly not all circular. Most of these intriguing patterns have appeared in fields, typically of wheat or barley, since the 1970s and documented cases have increased rapidly since then. However, there have been reports of crop circles spanning the last few hundred years. In 1678 it was suggested that a 'mowing devil' had been at work in a field in Hertfordshire, while John Rand Capron, an amateur scientist, wrote a letter to the editor of *Nature* in 1880 describing how a recent storm had created several circles of flattened crops in a field. Interestingly, in the case of the Hertfordshire circles, the stalks were flattened rather than broken – the same as in modern crop circles.

These formations are by no means restricted to the North Wessex Downs: they have been recorded in many countries, but the vast majority of them, about 90 per cent, are found in southern England and particularly in Wiltshire. Many of the formations, especially those in Wiltshire, appear close to ancient monuments, and according to one study, nearly half of all circles found in the UK in 2003 were located within a 9-mile radius of Avebury.

Since the early 1990s crop formations have increased in both size and complexity, some using as many as 2,000 different shapes, while others depict complex mathematical and scientific characters. In 1996, in a field near Windmill Hill, 194 crop circles appeared overnight, forming a complex pattern derived from an equation developed by Gaston Julia. This consisted of circles that defined three intertwined fractals, known as a 'triple Julia'. In 2001 a complex crop circle in the form of a double (six-sided) triskelion – a triskelion being a motif consisting of three interlocked spirals – measuring 780 feet in diameter and made from 409 circles, appeared in a field at Milk Hill near Alton Barnes.

In 2009 a 150-foot-long dragonfly formation appeared at Yatesbury, while near Wayland's Smithy in the same year a 600-foot-long giant jellyfish-like crop formation was seen. Other crop circles depict such things as the Mayan calendar; another was in the form of a coded image representing the first 10 digits (3.141592654) of 'pi' based on ten angular segments in concentric rings; some even show 3D effects when viewed from above.

In 2009 another formation appeared over the space of a week on Milk Hill in three separate stages, each stage adding more complexity. The first stage was said to show an 'astronomical sextant', the next night added more detail to the planets' orbits and the third brought the addition of strange hieroglyphs stretching across the field for 550 yards. As to what it all means, well that's still being debated, though some believe it was relating to a specific alignment of planets.

As to their origin, speculation has raged since they were first reported a few hundred years ago …

Some dismiss crop circles as utter rubbish, but despite years of research nobody knows conclusively how some of them are made. What we do know is that some are man-made; in 1991 two hoaxers, Douglas Bower and David Chorley, confessed to making 250 formations throughout England. Traditional circle-makers used 'stompers' – wooden planks attached to two hand-held ropes, which are a surprisingly efficient tool for flattening crops. However, modern designs have evolved beyond the traditional requirement that stalks be flattened rather than broken: formations now feature stalks that are carefully placed to create intricate textures within the shape.

And, as for how these people map the complex shapes and patterns onto the ground, in a field of knee-high wheat or barley, at night, let alone make them in the space of a few hours, without being seen, is definitely a mystery waiting to be answered.

One thing that can be said is that crop formations appear to be getting both larger and more complex in their design.

So maybe there are other forces at work; maybe crop formations are a form of communication, or are they the work of the 'mowing devil'? Maybe one day we'll know more, but for now, just try keeping an open mind on the subject.

CHAPTER SIX

Writers and artists

The isolated openness of the rural landscape, the beauty of the natural world and the intriguing ancient history of the North Wessex Downs have influenced and inspired many a writer and artist.

Paul Nash was captivated by the Wittenham Clumps and Silbury Hill while Sir Stanley Spencer painted stunning murals in a chapel at Burghclere relating to his experiences during the First World War. Richard Jefferies and Alfred Williams gave vivid portrayals of life in North Wiltshire, Richard Adams followed a group of rabbits in *Watership Down*, and Thomas Hardy set his novels in the semi-fictional county of Wessex, drawing on local towns and villages for inspiration.

The Bloomsbury Set

The Bloomsbury Set, often known as the Bloomsbury Group, was an influential group of writers, philosophers and artists that included such icons as Virginia Woolf and E. M. Forster.

The sixteenth-century Mill House in Tidmarsh on the River Pang became the country retreat of the Bloomsbury Set artist Dora Carrington (1893–1932) – later she was always known simply as Carrington – and author Lytton Strachey (1880–1932). The pair moved into the mill in 1917 around the same time that Lytton was publishing *Eminent Victorians*, the book, consisting of biographies of four leading Victorian figures – Cardinal Manning, Florence Nightingale, Thomas Arnold and General Gordon – that made him famous. Carrington painted many views of the house and surrounding countryside and one of her paintings of the mill adorns the front cover of *Carrington: Letters and Extracts from Her Diaries* published in 1979.

Carrington and Lytton both fell in love with ex-army officer Reginald Partridge (1894–1960), whom they renamed Ralph, forming à ménage à trois and in 1921, although Carrington loved the homosexual Lytton, Ralph and Carrington decided to marry. Shortly afterwards Ralph fell in love with the writer and diarist Frances Marshall (1900–2004), and in 1924 the group of four moved to Ham Spray House, to the south of Hungerford.

Strachey died suddenly at Ham Spray in 1932 and six weeks later Carrington committed suicide: following these two deaths, Ralph went on to marry his lover, Frances. The Partridges lived at Ham Spray House until 1961 when Frances sold the property following the death of Ralph.

During the group's lifetime, the home, surrounded by fields and extensive views, was well suited to both their communal living and working arrangements, and Frances wrote in her diary 'We believed there was no view more beautiful, more inexhaustible in England, and no house more lovable than Ham Spray.' Throughout her diaries and memoirs she often mentions walks along the Kennet and Avon Canal, through Savernake Forest and around the downs.

Carrington, who attended the Slade School of Art at University College, London, was, according to the former Director of the Tate Gallery, Sir John Rothenstein, 'the most neglected serious painter of her time'. These days her work is taken more seriously and two of her paintings – *Farm at Watendlath* (1921) and *Spanish Landscape with Mountains* (c.1924) – are now held by the Tate Gallery. Her hopeless love for Lytton Strachey was dramatized in the 1995 film *Carrington*.

Writers of Prose and Poetry

We have already read that G. K. Chesterton was in awe of the Uffington White Horse in his *Ballad of the White*

Horse, while the historical novelist Sir Walter Scott referred to the legend of Wayland's Smithy in his novel *Kenilworth*. However, over the last few hundred years many a writer has visited, lived in, or been inspired by, the North Wessex Downs. Some key figures are mentioned below, in approximate order of their periods of activity.

An early visitor was Jonathan Swift (1667–1745), Dean of St Patrick's Cathedral in Dublin, satirist, poet and author of *Gulliver's Travels*, who stayed at the seventeenth-century Old Rectory in Letcombe Bassett in 1714 where Alexander Pope came to visit him before Swift left for Ireland.

The poet Stephen Duck (1705–1756) was born into a poor labouring family at Charlton St Peter to the south-west of Pewsey, right on the edge of the AONB. Leaving school at 14, he followed the family and started working in the fields. Yet following his first marriage he started reading poetry and literature in an attempt to better himself and in 1730 he wrote *The Thresher's Labour*, a poem that portrays the difficulties of an agricultural life. The poem was celebrated throughout London society, bringing him to the attention of Queen Caroline, who became his patron.

William Cobbett (1763–1835), writer, agriculturist and political reformer, born in Farnham, Surrey, is best known for his book *Rural Rides*. He originally worked as a farm labourer before moving to London, after which followed several years in the army, before he returned to England in 1799; three years later he started his newspaper *The Political Register*. After a two-year period of self-imposed exile in America, Cobbett returned with a determination to both set about reforming Parliament and helping to put an end to poverty for farm labourers.

With this in mind, in 1821, Cobbett set off on a series of journeys on horseback through the countryside of Southern England with the objective 'not to see inns and turnpike-roads, but to see the country: to see the farmers at home, and to see the labourers in the fields', writing down what he encountered from the points of view both of a farmer and a social reformer. His journeys through the countryside were originally printed in *The Political Register* between 1822 and 1826 before he later combined them to form the basis of *Rural Rides*, which was published in 1830.

On one of his journeys he takes us through 'Auborne' (or Aldbourne) to Ramsbury 'a large, and, apparently, miserable village, or "town" as the people call it' before reaching Shalbourne, of which he writes: 'Shallburn, where, Tull, the father of the drill-husbandry, began and practised that husbandry at a farm called "Prosperous"'.

Cobbett's journey continues through Burghclere, passing Lord Carnarvon's estate at Highclere and Beacon Hill before arriving at Hurtsbourne Tarrant, situated along the southern edge of the AONB. Cobbett often visited Joseph Blount, who lived at Rookery Farm House, situated at the base of The Hill in Hurstbourne Tarrant (now the A343); the wall of the front garden was known as the 'Wayfarers' Table', because the farmer used to leave food for passing travellers and hungry farm labourers. Cobbett writes:

> At Uphusband [we are told elsewhere in the book that Uphusband is in fact Hurstbourne Tarrant], a little village in a deep dale, about five miles to the North of Andover, and about three miles to the South of the Hills at Highclere. The wheat is sown here, and up, and, as usual, at this time of the year, looks very beautiful. The wages of the labourers brought down to six shillings a week! a horrible thing to think of; but, I hear, it is still worse in Wiltshire.

Another journey took him to the Vale of Pewsey in the far south-west of the North Wessex Downs:

> From the top of a very high part of the down, with a steep slope towards the valley, I first saw this Valley of the Avon; and a most beautiful sight it was! Villages, hamlets, large farms, towers, steeples, fields, meadows, orchards, and very fine timber trees, scattered all over the valley … I sat upon my horse and looked over Milton and Easton and Pewsey for half an hour … I never before saw anything to please me like this valley of the Avon.

Rural Rides is still in print and gives a fascinating insight to the lives of those living and working in the countryside during the early nineteenth century.

Lawyer and author Thomas Hughes (1822–1895) was born at Uffington in the Vale of White Horse just to the north of the North Wessex Downs and spent his childhood years in the village. His most famous book was *Tom Brown's Schooldays* (1856), in which the little village school is mentioned, and there can be little doubt that the adventures of Tom Brown are closely modelled on the author's own childhood (the sequel, *Tom Brown at Oxford*, was written in 1861).

In *Tom Brown's School Days* Hughes writes of Whitehorse Hill: 'and then what a hill is the White Horse Hill! There it stands right above all the rest, nine hundred feet above the sea, and the boldest, bravest shape for a chalk hill that you ever saw... a place that you won't ever forget.'

Hughes also wrote *The Scouring of the White Horse* (1859) which gave an important insight to the joyous festivities that accompanied the scouring of the famous Uffington White Horse in 1857.

The old village school that Hughes attended is now home to the Tom Brown's School Museum, which not only has mementoes of Thomas Hughes and the late poet laureate, Sir John Betjeman, who also lived in Uffington for several years, but also explains the history and archaeology of the area.

The unusual twin tops of the Wittenham Clumps – Castle Hill and Round Hill – crowned by stands of beech trees, have inspired both writers and artists and offer some stunning views. Castle Hill, also crowned with the remains of an Iron Age hill fort, hence its name, was also home to the Poem Tree. Here, on a beech tree, Joseph Tubb (1805–1875) of Warborough Green carved a poem in 1844–5 inspired by his love of the surrounding landscape. Sadly the old beech tree finally collapsed in the summer of 2012; however, on the 150th anniversary of the carving a memorial stone was placed close to the trees, with an inscription of the poem taken from a tracing made by Dr Henry Osmaston in 1965:

As up the hill with labr'ing steps we tread
Where the twin Clumps their sheltering branches spread
The summit gain'd at ease reclining lay
And all around the wide spread scene survey
Point out each object and instructive tell
The various changes that the land befell
Where the low bank the country wide surrounds
That ancient earthwork form'd old Mercia's bounds
In misty distance see the barrow heave
There lies forgotten lonely Cwichelm's grave.

Around this hill the ruthless Danes intrenched
And these fair plains with gory slaughter drench'd
While at our feet where stands that stately tower
In days gone by up rose the Roman power
And yonder, there where Thames smooth waters glide
In later days appeared monastic pride.
Within that field where lies the grazing herd
Huge walls were found, some coffins disinter'd
Such is the course of time, the wreck which fate
And awful doom award the earthly great.

Much earlier another poet, Matthew Prior (1664–1721), had also been inspired by the Wittenham Clumps, for this is where he is said to have written his poem *Henry and Emma*, while sitting under an oak tree.

Novelist and poet Thomas Hardy (1840–1928), born in Dorset, set his fictional works that explored tragic characters struggling against their passions and social circumstances, in a Wessex that was semi-fictional, although based on the medieval Anglo-Saxon kingdom of Wessex which includes parts of Wiltshire, Hampshire and Berkshire that form the North Wessex Downs AONB.

Hardy knew the area well, using features and places as the basis for parts of his novels such as the remains of a Neolithic burial chamber known as the Devil's Den in his short story, 'What the Shepherd Saw' (1881). His last, and some claim most controversial, novel, *Jude the Obscure* (1895), was set in north Wessex at the fictional village of 'Cresscombe', based on Letcombe Bassett; a beautiful old thatched cottage next to the stream is said to have been the inspiration for Arabella's cottage. Other

▲ Plaque on Liddington Hill commemorating two local writers: Richard Jefferies and Alfred Williams (see pages 149–50)

places used for inspiration were Wantage ('Alfredston'), Fawley ('Marygreen'), Reading ('Aldbrickham') and Newbury ('Kennetbridge'). During Jude's short quest to reach 'Christminster' (Oxford) he followed 'a white road [that] seemed to ascend and diminish till it joined the sky. At the very top it was crossed at right angles by a green "ridgeway" – the Icknield Street and original Roman road through the district.'

On the north wall of the nave at the Church of St Peter and St Paul in Yattendon is a tablet commemorating Harriet Molesworth, her son Robert Seymour Bridges (1844–1930) and his wife, Monica Waterhouse. Bridges practised medicine until his retirement on health grounds in 1881 and following his marriage in 1884 he lived in the village of Yattendon to the northeast of Newbury (his ashes are buried in the churchyard close to the family cross). Although not a well-known poet, Bridges became poet laureate in 1913. His works include *The Testament of Beauty* (1929), a long philosophical poem which sought to reconcile scientific knowledge with Christian faith, and the popular *London Snow* (1879), of which the following is an extract:

When men were all asleep the snow came flying,
In large white flakes falling on the city brown,
Stealthily and perpetually settling and loosely lying,
Hushing the latest traffic of the drowsy town;
Deadening, muffling, stifling its murmurs failing;
Lazily and incessantly floating down and down:
Silently sifting and veiling road, roof and railing;
Hiding difference, making unevenness even,
Into angles and crevices softly drifting and sailing.

Passionate about the beauty of the countryside and richness of nature that he saw all around him, the Victorian writer Richard Jefferies (1848–1887) became particularly noted for his depiction of English rural life. The son of a farmer, Jefferies was born in the small village of Coate near Swindon, just to the north of the AONB in 1848. The old farmhouse where he grew up is now a museum dedicated to his life and works; his book *Bevis – the Story of a Boy* drew on his boyhood adventures with his brother while he was living at Coate.

In 1866 Jefferies started working as a journalist for several local newspapers and it was William Morris, editor of the *Swindon Advertiser*, who encouraged him to write more. He started writing articles on local history and then wrote *Reporting, Editing, and Authorship: Practical Hints for Beginners in Literature* (1873), in which he shared the knowledge he had gained as a local reporter. In 1874, the year he married Jessie Baden, his first novel, *The Scarlet Shawl*, was published.

After the birth of their baby son, Richard and Jessie moved close to Surbiton and he spent much of his time walking through the countryside. These walks later provided the material for *Nature near London* published in 1883.

His articles on rural life in Wiltshire for the *Pall Mall Gazette* were well received and a collection of these formed the basis of *The Gamekeeper at Home* (1878). Following publication, Jefferies started to be compared with the great English nature writer, Gilbert White. Three more collections of essays that had appeared in the *Pall Mall Gazette* followed: *Wildlife in a Southern County* (1879), *The Amateur Poacher* (1879), and *Round About a Great Estate* (1880). A fourth collection, *Hodge and his Masters* (1880), brought together articles that had first been published in the *Wilts and Gloucestershire Standard*.

In *Wildlife in a Southern County* he gives an evocative description of the Ridgeway, which still holds true today:

> A broad green track runs for many a long, long mile across the downs, now following the ridges, now winding past at the foot of a grassy slope, then stretching away through a cornfield and fallow. It is distinct from the wagon-tracks which cross it here and there, for these are local only, and, if traced up, land the wayfarer presently in a maze of fields, or end abruptly in the rickyard of a lone farmhouse. It is distinct from the hard roads of modern construction which also at wide intervals cross its course, dusty and glaringly white in the sunshine … with varying width, from twenty to fifty yards, it runs like a green ribbon.

However, his health was failing; he had been suffering from tuberculosis for a number of years, and in 1883 he published his thought-provoking autobiography, *The Story of My Heart*.

His next novel, *After London* (1885), was a fictional look at life in the aftermath of some unspecified catastrophic event that had drastically reduced the human population; the land reverted back to nature, and the few people who had survived fell back on a medieval way of life. His final novel was *Amaryllis at the Fair* (1887); most probably based on his own upbringing, the novel paints a picture of the ups and downs of life on a small, family-run farm burdened by debt.

He moved to Goring-by-Sea (Worthing) in the hope that the sea air might aid his health, but the disease that had been with him for so many years finally took its toll in 1887; he is buried in Broadwater Cemetery, Worthing. His widow produced *Field and Hedgerow; Being the Last Essays of Richard Jefferies* in 1889; a collection of essays that Jefferies had planned to publish before his untimely death.

Born in Edinburgh, Kenneth Grahame (1859–1932) came to live with his grandmother at Cookham Dean (Berkshire) at the age of five following the death of his mother. After his retirement from his job as Secretary at the Bank of England in 1908, Grahame moved back to Cookham. He later moved to Bonham's Farmhouse in Blewbury from 1910 until 1920 and then to Church Cottage next to St James' Church in Pangbourne from 1924 until his death. Grahame is best known for his children's classic, *The Wind in the Willows* (1908), which was based on the bedtime stories about animal characters living beside a river that he used to tell to his young son, Alastair (the stories were inspired by his time spent at Cookham Dean).

However, in his first book, *Pagan Papers* (1893), a wide-ranging collection of essays, he describes the Ridgeway and surrounding landscape in the following terms:

> The best example I know of an approach to this excellent sort of vitality in roads is the Ridgeway of the North Berkshire Downs. Join it at Streatley, the point where it crosses the Thames; at once it strikes you out and away from the habitable world in a splendid purposeful manner, running along the highest ridge of the Downs, a broad green ribbon of turf, with but a shade of difference from the neighbouring grass, yet distinct for all that. No villages nor homesteads tempt it aside or modify its course for a yard; … Out on that almost trackless expanse of billowy Downs such a track is in some sort humanly companionable: it really seems to lead you by the hand.

After graduating, Laurence Binyon (1869–1943) worked at the British Museum and in 1915, despite being too old to enlist in the First World War, he started working at a British hospital for French soldiers, the Hôpital Temporaire d'Arc-en-Barrois, Haute-Marne, in northeastern France. He returned in the summer of 1916 and took care of soldiers taken in from the Verdun battlefield. He wrote about his experiences in *For Dauntless France* (1918) and his poems, *Fetching the Wounded* and *The Distant Guns*, were inspired by his time spent working at the hospital.

Yet it is his poem *For the Fallen*, written in 1914 at the start of the First World War, for which he is most remembered. His immortal lines are inscribed on Lutyen's Cenotaph in Whitehall, London, and recited at Remembrance Day services:

> *They shall not grow old, as we that are left grow old:*
> *Age shall not weary them, nor the years condemn.*
> *At the going down of the sun and in the morning,*
> *We will remember them.*

Following his retirement in 1933, Laurence and his wife, Cicely, went to live at Westridge Farm just to the east of Aldworth. He continued to publish poems, including *The North Star and Other Poems* (1941) and *The Burning of the Leaves and Other Poems* (1944), and at the time of his death he was working on a major three-part Arthurian trilogy, the first part of which was later published as *The Madness of Merlin* (1947); his ashes are buried in St Mary's churchyard, Aldworth.

Alfred Williams (1877–1930) lived at South Marston, near Swindon, not far from where Richard Jefferies once lived. He grew up in a life of poverty; his father had left home when he was five and his mother had to raise the children single-handed whilst trying to make ends meet. His education was poor and by the age of eight he was already working part of the day on a farm, while the rest of the time was spent at school. Three years later he left school to become a full-time farm labourer, which suited his love of the countryside. Yet the draw of better pay persuaded him to follow two of his brothers and go 'inside', working at the Great Western Railway factory in Swindon from the age of fifteen (people in Swindon always called working at the GWR factories being 'inside').

Although Williams worked in the steam-hammer shop at Swindon Railway Works, his heart and soul remained in his beloved countryside, of which he was a keen observer: 'The slopes of the downs, if they have general forms, are continually changing and interchanging in localities, assuming new and strange shapes, charming and surprising with their grace and exquisiteness … for ever reflecting the mood of the heavens.'

In his spare time he continued his studies, read the Classics, tried his hand at painting, learnt languages and generally absorbed the everyday life around him. By 1907 he was writing his own verse and two years later he published his first of six collections of poetry – *Songs of Wiltshire* (1909), quickly followed by *Poems in Wiltshire* in 1911. The following is an extract from *Liddington Hill* published in *Songs in Wiltshire*:

The friendship of a hill I know
Above the rising down,
Where the balmy southern breezes blow
But a mile or two from town;
The budded broom and heather
Are wedded on its breast,
And I love to wander thither
When the sun is in the West.

Dubbed 'the hammerman poet', throughout his life he was to write both poetry and prose that showed his deep love of the countryside around him, but his books were never commercial enough to allow him to leave the factory. In fact, what is probably his greatest work was *Life in a Railway Factory*, recounting his experiences at the Swindon Railway Works. The book, written while he still worked for GWR, could not be published until he had left their employment; the grim picture of the life endured by the workers that the book portrayed would most likely have led to his immediate dismissal by the company. The book was finally published in 1915 shortly after Williams had to leave the works on health grounds, yet whilst still working at the factory he continued to write, publishing *A Wiltshire Village* (1912) and *Villages of the White Horse* (1913).

When *Life in a Railway Factory* was finally published it immediately won critical acclaim; *The Daily Chronicle* (a leading newspaper of the day) called it 'a book of revelation', while *The Times* commented that Williams was 'a born observer with a gift of words and a love of truth'.

By now the First World War was taking its dreadful toll and Alfred, despite his ill-health, volunteered for service. He became a gunner in the Royal Field Artillery and served in India, a country which immediately fascinated him. After the war Williams returned to a life of poverty, but he continued to write, publishing *Round about the Upper Thames* (1922) and *Folk Songs of the Upper Thames* (1923). He and his wife, Mary, built a small cottage that they called *Ranikhet* – named after the hill station in India, within sight of Mount Everest where he served in 1917. However, life was very difficult and after Mary was diagnosed with cancer, Alfred in 1930 died of a broken heart. His last work, an English translation of the classic Sanskrit collection of India fables called *Tales from the Panchatantra*, was published shortly after his death.

A memorial stone to both Alfred Williams and John Jefferies is located high up on Burderop Down close to Barbury Castle (SU158672), an area much loved by both writers and a fitting place for a memorial (see page 138).

Another poet involved in the First World War was Edward Thomas (1878–1917), who lived at Steep in Hampshire. He was killed in action at the Battle of Arras, on the western front, on Easter Monday 1917. Following his death his wife, Helen, moved to Eastbury, where she remained until her death in 1967. The local Church of St James the Greater in Eastbury has a fitting tribute to the poet and his wife, a stunningly detailed engraved glass window by Sir Laurence Whistler (see Interesting Churches – page 130).

Thomas loved to wander through the rolling contours of the Hampshire and Berkshire Downs and his writing vividly describes the landscape. He once set out to trace the prehistoric Icknield Way and write a book about it (*The Icknield Way* – 1913), spending time studying in the British Museum before walking much of the way. Part of his journey took him from Streatley past Scutchamer Knob, the Devil's Punchbowl, Whitehorse Hill and Wayland's Smithy.

Of the Icknield Way he writes:

At Uffington Castle it is over eight hundred feet high, but a little lower than the highest part of the camp. From the rampart about this circle of almost level turf I could see the Quarley Hill range and far over the Lambourn Downs to Martinsell Hill by Savernake; I could see Barbury Castle and the wooded hills of Clyffe and Wroughton, and Badbury, the Cotswolds, the Oxfordshire hills, Sinodun, and the Chilterns. The Dragon Hill below it is an isolated eminence shaped like the butt of an oak tree … the Ridgeway went fairly straight, with a thorn or two on either side, towards the thick beech clump above Wayland's Smithy, sometimes a green road, sometimes worn white.…the Ridgeway is like nothing so much as a

▲ Looking eastwards towards Uffington Castle from the Ridgeway

battlement walk of superhuman majesty. The hills between Streatley and Liddington form a curve in the shape of a bow, a doubly curved Cupid's bow. Following this line, always keeping at the edge of the steep north-ward slope and surveying the valley, the Ridgeway carries the traveller for thirty miles as if along the battlements of a castle.

Thomas also wrote *Richard Jefferies: His Life and Work* (1909), a biography of a man (mentioned earlier) whose own writings had a profound influence on him. In it, Thomas captures the sense of remoteness and huge vastness of the Berkshire Downs:

> The Downs in this immediate country of Jefferies are among the highest, most spacious, and most divinely carved in rolling ridge and hollow flank; and their summits commune with the finest summits of the more southerly Downs – Inkpen, Martinsell, Tan Hill … Jefferies often thought of the sea upon these hills. The eye sometimes expects it. There is something oceanic in their magnitude, their ease, their solitude above all, in their liquid forms, that combine apparent mobility with placidity, and in the vast playground which they provide for the shadows of the clouds. They are never abrupt, but, flowing on and on, make a type of infinity … a hugeness of undivided surface for which there is no comparison to be found on the earth, and but seldom in the sky.

Charles Hamilton Sorley (1895–1915) was a contemporary of Edward Thomas, and like Thomas was a victim of the First World War. Sorley, who attended Marlborough College, wrote a number of poems inspired by the varied landscape of the downs. The following is an extract from his poem 'Marlborough' (1914):

> *Crouched where the open upland billows down*
> *Into the valley where the river flows,*
> *She is as any other country town …*
>
> *… I, who have walked along her downs in dreams,*
> *And known her tenderness and felt her might,*
> *And sometimes by her meadows and her streams*
> *Have drunk deep-storied secrets of delight …*

In 'East Kennet Church at Evening' (1913) he wrote:

I stood amongst the corn, and watched
The evening coming down.
The rising vale was like a queen,
And the dim church her crown.

While the Iron Age hill fort at Barbury inspired him to write 'Barbury Camp' (1913):

We burrowed night and day with tools of lead,
Heaped the bank up and cast it in a ring
And hurled the earth above. And Caesar said,
Why, it is excellent. I like the thing.

The novelist and poet D. H. Lawrence (1885–1930) and his German wife Frieda moved to Chapel Farm Cottage, Hermitage, in 1917. Lawrence, author of such works as *Lady Chatterley's Lover* (1928), spent two years at the cottage and the novel most closely linked with his time there was *The Fox* (1923). The book was set at the fictional Bailey Farm, believed to have been based on Grimsbury Farm in Long Lane. The setting of the farm is recreated in the book, while the nearby market town is likely to have been Newbury.

The former poet laureate Sir John Betjeman (1906–1984) and his wife, Penelope, lived at Garrard's Farm (blue plaque) in the village of Uffington from 1934 to 1941 before moving to the Old Rectory at Farnborough in 1945. Penelope wrote to a friend that year 'Father has bought us a beautiful William and Mary house 700 feet up on the downs above Wantage, with 12 acres of land, including a wood and two fields. It is a dream of beauty but has no water and no light and is falling down and needs six servants, so it will probably kill us in the end.' The village church – All Saints – contains a beautiful memorial window to Betjeman by his friend John Piper (see the Interesting Churches section on page 131).

Penelope Betjeman loved to explore the downs on horseback and her daughter Candida placed a sarsen stone beside the Ridgeway in her memory. John preferred to wander through the rolling countryside on foot, catching glimpses of the varied life around him, as evidenced by these extracts from his poem 'Upper Lambourne':

Where, in nineteen-twenty-three,
He who trained a hundred winners,
Paid the Final Entrance Fee …

Feathery ash in leathery Lambourne
Waves above the sarsen stone …

Sir William Gerald Golding (1911–1993), the novelist, poet and playwright, who was awarded the Nobel Prize in Literature in 1983, is best known for his first novel, *Lord of the Flies* (1954). Although born at his grandmother's house in Cornwall, where he spent many childhood holidays, William Golding grew up at the family home in Marlborough, where his father was a science master at Marlborough Grammar School.

He renamed Marlborough as *Stilbourne* and made it the setting for his 1967 novel *The Pyramid*, 'beset with class distinctions and repressed desires', while in an essay on Wiltshire he wrote that the land 'is seamed and furrowed with ditches, erupts with grassy forts and is scattered with the mounds of enigmatic graves'.

The novelist Richard Adams (b.1920) is best known as the author of *Watership Down*, his modern classic published in 1972 about a group of rabbits who struggle to survive against all the odds. Adams had originally begun telling the story of *Watership Down* to his two daughters, before they persuaded him to publish it as a book.

Although he now lives at Whitchurch, Adams grew up at Wash Common to the south-west of Newbury and in his autobiography, *The Day Gone By,* he describes the view from his childhood home:

The superb view to the south was across the open country of ploughland, meadows and copses typical of the Berkshire-Hampshire border, stretching away four or five miles to the distant line of the Hampshire Downs – the steep escarpment formed by Cottington's Hill, Cannon Heath Down, Watership Down and Ladle Hill.

Places mentioned in *Watership Down* are all located within easy reach of where Adams grew up. Close by was Sandleford, the location of the warren where Hazel, Fiver and Bigwig began life; nearby was the River Enborne which the rabbits had to cross to make their escape, while visible on the horizon was their new home at Watership Down. Sadly, fiction has become reality for 'Sandleford Warren' following the decision to allow the future development of new housing at Sandleford Park, although part will be set aside as a country park.

Artists of Land and Light

The legendary sites of the North Wessex Downs have been depicted over the years by many an artist, such as Eric Ravilious (1903–1944), who painted the *White Horse at Uffington* as part of a series of chalk hill figures intended as illustrations for a children's book that unfortunately was never completed.

Throughout his artistic career, the English landscape painter Paul Nash (1889–1946) was drawn to places in the landscape with ancient, mystical connections which linked with his belief in the *genius loci*, or spirit, of certain landscapes. Two such places were the Avebury stone circle and the Wittenham Clumps which featured in his paintings including *Landscape on a Vernal Equinox* (1944).

Nash, a talented writer, book illustrator and surrealist, was also an official war artist – his paintings of First World War trenches vividly portray the destruction, while during the Second World War he created iconic works such as *Totes Meer* (German for 'dead sea'), depicting a landscape littered with wrecked German aircraft. When, in 1911, Nash first encountered the tree-crowned Wittenham Clumps tucked in the far north-eastern corner of the AONB, he was immediately caught by their striking outline and mystical presence and wrote of his discovery, calling it 'a beautiful legendary country'. Throughout his life, since he first painted them in 1912, they became a source of inspiration to which he returned on many an occasion.

The prehistoric remains at Avebury were ideal subjects for his imaginative style of painting, allowing him to make them objects of mystery in his *Landscape of the Megaliths* series. He wrote of Avebury:

The great stones were then in their wild state, so to speak. Some were half covered by the grass, others stood up in the cornfields or were entangled and overgrown in the copses, some were buried under the turf. But they were always wonderful and disquieting, and, as I saw them, I shall always remember them … their colouring and pattern, their patina of golden lichen, all enhanced their strange forms and mystical significance.

Sir Stanley Spencer (1891–1959) was a famous painter who, for most of his life, lived in the Berkshire village of Cookham on the banks of the River Thames to the east of the AONB, and is noted for his works depicting Biblical scenes. Many of his works were set in the village of Cookham and featured local residents, such as in *The Resurrection, Cookham* (1923–7). The former Wesleyan Chapel where Spencer used to worship as a child now houses the Stanley Spencer Gallery.

Spencer served with the Royal Medical Army Corps and the Berkshire Regiment during the First World War and his experiences greatly influenced his work. He later worked as a war artist during the Second World War.

One of his major commissions, which he completed in 1932, was for the Sandham Memorial Chapel in Burghclere a couple of miles east of Highclere Castle. The unassuming red-brick chapel built in memory of Lieutenant Sandham, who died in the First World War, houses a stunning collection of First World War murals inspired by Spencer's own experiences, in honour of the 'forgotten dead' of the First World War, who were not remembered on any official memorials.

The interior is dominated by the *Day of Resurrection* scene behind the altar, in which British soldiers lay the white wooden crosses that had marked their graves at the feet of Christ. The series, which chronicles Spencer's everyday experiences of the war, rather than any scenes of action, is considered to be one of his finest works, leading some to suggest that they are 'Britain's answer to the Sistine Chapel'. While working on the murals he found time to undertake some paintings of the local area, including *Beacon Hill near Highclere* (1927) and *Cottages at Burghclere* (1930).

Eric Ennion (1900–1981), born in Northamptonshire, the son of a country doctor, studied medicine before joining his father's practice, yet from an early age he was fascinated by watching and drawing birds. At the end of the Second World War he sold the medical practice and moved to Flatford Mill with his wife Dorothy; they later moved to Northumberland where they established a field centre and bird observatory. In 1961 they moved to the Mill House in Shalbourne, from where he continued to paint and run courses on wildlife and landscape painting until his death; during the early years at Shalbourne he wrote a series of fortnightly articles, 'The Countryman's Log', for *Farm and Country* magazine. Throughout his career, Ennion's paintings have illustrated many books, including some that he wrote himself, such as *The House on the Shore* (1960) about his time in Northumberland.

John Piper (1903–1993), who lived for over fifty years at Fawley Bottom in Buckinghamshire, near Henley-on-Thames, was much-inspired by the North Wessex Downs, including Avebury, which featured in such paintings as *Avebury* (1936) and *Avebury Restored*. However, he is probably best known for designing the stained-glass windows in Coventry Cathedral. Within the North Wessex Downs he designed a stained-glass window for All Saints Church in Farnborough in memory of his good friend Sir John Betjeman. In 1981 he was commissioned to design a stained-glass window for the Wiltshire Heritage Museum in Devizes, which can be seen there. This window, interpreted by Patrick Reyntiens, shows several features of the surrounding landscape: the Ridgeway, the Devil's Den on the Marlborough Downs, the stones of the West Kennet Avenue at Avebury, the Iron Age hill fort of Barbury Castle, the Cherhill White Horse and the woolly headed thistle, as well as featuring some archaeological finds including the Upton Lovell amber necklace and Bronze Age pottery.

The landscape artist Nick Schlee (b.1931) is known for his dynamic oil paintings and woodcuts of Wiltshire, Oxfordshire, Hampshire and Berkshire, including the ancient stone monuments of Avebury and the Ridgeway.

Famed for his painting *The Badminton Game*, the artist David Inshaw (b.1943) moved to Devizes in 1971 and formed the Broadheath Brotherhood with Graham and Ann Arnold in 1972. These three artists were joined by Peter Blake, Jann Haworth, and Graham and Annie Ovenden in 1975, and the group was renamed the Brotherhood of Ruralists; they exhibited together for the first time at the Royal Academy Summer Exhibition in 1976. Inshaw left the group in 1983 and after living in Hay-on-Wye for several years he moved back to Devizes in 1995.

Inshaw fell in love with the landscape of the North Wessex Downs; the Marlborough Downs and local wildlife all made frequent appearances in his work. However, his most symbolic feature was the Neolithic Silbury Hill which appeared in many of his paintings, including *Silbury from the Air* (1988), *Portrait of Silbury Hill* (1989), *Storm over Silbury Hill* and *Silbury Hill in the Moonlight* (both 2008).

Anna Dillon (b.1972) is a modern semi-abstract landscape painter; born at Wallingford, Anna grew up in both Wiltshire and Oxfordshire close to such inspiring places as Avebury, the Uffington White Horse and the Ridgeway. This unique and ancient countryside inspires her to paint vibrant and colourful landscapes. In 2012 she completed a commission from the North Wessex Downs AONB Partnership to illustrate ten of the most iconic features within the North Wessex Downs, the subject matter being the Avebury stone circle, the stark outline of the Combe Gibbet, The Kennet and Avon Canal, a lapwing, racehorse training on the downs, Savernake Forest, The Uffington White Horse, a rabbit on Watership Down, Wilton Windmill and the tree-crowned Wittenham Clumps.

Split into series such as Wiltshire, Oxfordshire, and the Ridgeway, Anna's paintings vividly portray many parts of the North Wessex Downs countryside including The Devil's Punchbowl and the West Kennet Avenue at Avebury.

CHAPTER SEVEN

Wildlife

The North Wessex Downs has a rich and varied wildlife owing to the diverse range of land types and the relative isolation of the region, and these factors combine to make the region an ideal place in which to catch glimpses of wildlife at fairly close quarters. Take a walk along quiet autumnal lanes and the chances are that you'll be accompanied by flocks of redwing and fieldfare feeding on the autumn berries; as dusk approaches, deer come out from the cover of trees to feed; on summer days, you'll frequently hear skylarks singing, or see the majestic outline of a buzzard soaring effortlessly high above, while spring and summer bring forth a myriad of flowers and butterflies.

The Flora and Fauna

ARABLE FARMLAND

The dominant land use within the North Wessex Downs is large-scale open arable farming, providing a habitat for a range of farmland birds and mammals. Up until the middle of the twentieth century, arable fields would have contained a much greater number of plants and flowers than just wheat and barley. Fortunately an increase in environmental awareness and wildlife-friendly farming schemes has led to farmers leaving wider field margins and setting aside areas to allow these plants to thrive, creating important areas for wildlife that not only provide safe nesting habitats, but also provide insect-rich food sources during the summer and seed-rich food sources during the winter.

Populations of farmland birds across the UK have shown a dramatic decline over recent decades; some, such as the turtle dove, have declined by over 90 per cent. Ten species – namely the corn bunting, grey partridge, lapwing, linnet, reed bunting, tree sparrow, turtle dove, skylark, yellow wagtail and yellowhammer – are of particular concern due to their populations showing the most severe declines. Fortunately the AONB has nationally important populations of these native farmland birds.

A rare and highly protected migratory summer visitor to the North Wessex Downs is the stone curlew (although not related to the common curlew, the stone curlew has a similar call). The RSPB have been working with farmers in the Wessex region for several decades to try and protect the species. This work has included the provision of suitable nesting plots so that the birds do not get disturbed by farm operations (although some birds still nest on the spring-tilled arable fields) and numbers are increasing; it is estimated that there are now about fifty pairs within the AONB (especially just to the north of Salisbury Plain).

Winter visitors include golden plover, brambling and flocks of fieldfare and redwing. Mammals include the brown hare (these are much bigger than rabbits with

▲ The North Hampshire Downs in winter, looking west from near Inkpen Hill

WILDLIFE

▲ Classic outline of a red kite; reintroduced in the Chilterns, these birds are now spreading into the North Wessex Downs

longer ears, usually tipped with black) and the harvest mouse, which feeds on berries and seeds.

Reintroduced into the Chilterns, having been driven to extinction by the end of the nineteenth century, red kites have spread westwards into parts of the North Wessex Downs, especially along the M4 corridor (the birds feed mostly on carrion). Red kites are easily identifiable by their distinctive forked tail, chestnut-red plumage and high-pitched whistling calls – 'weeoo-wee-weewee' – whereas buzzards have broader wings and a rounded tail.

CHALK GRASSLAND

Overlying the chalk bedrock are thin, well-drained, nutrient-poor soils that support a vegetation of herbs and grasses that has traditionally been grazed by sheep, cattle and rabbits. This is chalk grassland, one of the most distinctive and ecologically notable habitats of the North Wessex Downs. 'Unimproved' chalk grassland is one of the most biologically rich and diverse habitats in the UK – plant surveys have shown that over forty species of flowering plant can be found in a single square metre of chalk grassland – supporting a diverse community of invertebrates, mammals and birds.

Slightly less than 9 per cent of the UK's chalk grassland lies within the AONB; the area of chalk grassland was probably at its most extensive during the sixteenth century and has been declining ever since; in a fourteen-year period from 1966 the area of chalk grassland was reduced by 30 per cent and today only fragments of this habitat remain, as a consequence of a growing intensification of agriculture and the ploughing of the light, shallow downland soils with an associated decline in grazing.

These chalk grasslands provide food and breeding cover for birds such as skylarks and support important butterfly populations including the adonis blue, silver-studded blue, marsh fritillary, chalkhill blue, small blue,

▲ Sheep grazing on the chalk grassland below Whitehorse Hill

▲ Clustered bellflower: one of the many plants to be found on chalk grassland

silver-spotted skipper and Duke of Burgundy fritillary. The downland also supports a large number of scarce plant species including early gentian, eyebright, pasque flower, Chiltern gentian, dwarf mouse ear, tuberous thistle, field fleawort, round-headed rampion, burnt orchid, bastard toadflax and musk orchid.

It may seem strange, but grassland in Britain is not the natural vegetation; over time it would revert to woodland. The main reason this does not occur is down to land management: since prehistoric times, the grazing of animals has hindered the growth of new trees by preventing the saplings from becoming established. Grazing animals allow grasses, such as quaking grass, fescues, false oat and rye grass to become the dominant species. So long as the grass is kept short by grazing, a diverse range of plants can flourish, including salad burnet, wild thyme, rockrose, harebell, bird's foot trefoil, wild carrot, small scabious, clustered bellflower, milkwort, knapweed and several varieties of thistle, as well as many orchid species. Any reduction in grazing would allow scrub to invade, with the result that the ground would become shaded by taller plants and fewer species could thrive, thereby reducing the plant diversity.

WOODLAND

The North Wessex Downs has a significant concentration of ancient semi-natural broadleaved woodlands centred on the Berkshire and Marlborough Downs, the Hampshire Downs and in Savernake Forest. The diverse woodland types that make up these ancient woodlands include significant areas of wood pasture and support a wide range of species, with roosting sites for a number of bat species including Bechstein's, Barbastelle, greater horseshoe and noctule. Of particular importance are the calcareous woodlands, which support a range of rare plants including herb-paris and green hellebore and provide a home to a high proportion of the world's population of bluebells, which carpet the woodland floor in spring.

The type of woodland is based on the prevailing environmental conditions; however, tree species include oak, ash, birch, wych elm, hornbeam, hazel and beech. Beech woodlands are, however, limited in extent – especially when compared to the adjacent Chilterns AONB – and are restricted to beech hangers along the escarpments on the northern edge of the AONB and the escarpment along the northern edge of the Hampshire Downs. Along the valleys, such as the Kennet Valley, there are important examples of wet alder woodland.

The woods form important habitats for butterflies such as the pearl-bordered fritillary; hazel coppice provides important habitat for dormice, and rare plants include herb-paris and green hellebore. A large part of Savernake Forest has been designated as a SSSI in recognition of its outstanding lichen flora, fungi, rare invertebrates and breeding bird community.

CHALK STREAMS AND RIVERS

The spring-fed, fast-flowing chalk streams and rivers of the North Wessex Downs support a wide range of plant and animal life, with fish populations including perch, chub, grayling and brown trout: this last one is a native of Britain – rainbow trout were introduced from North America and only rarely breed in a natural British river environment.

Chalk streams such as the Kennet have been famous for trout fishing since the late nineteenth century; this form of fishing uses a 'dry-fly', which imitates a real insect or larva that will hopefully trick the fish into rising to catch it. The trout population is not just linked to the

▲ The River Kennet near Hungerford

▲ The Devil's Punchbowl, Hackpen Hill

quality and amount of water in the river, but also to the environment, which must supply plentiful hatches of appropriate flies for the fish to eat. Fortunately, these streams and rivers do support large numbers of insects including mayflies (the easy to identify long-tailed adults are a favourite food of trout), caddis flies and many species of water beetle.

Alongside the riverbanks, the reed beds and grasses are home to grass snakes and birds including reed buntings, as well as water voles. More recently otters have been making a welcome return to the streams and rivers of the North Wessex Downs.

Water plant life includes starwort, water crowfoot which forms floating carpets of white flowers in early summer, water forget-me-not, watercress and water speedwell. The wooded riverbanks, which are often seasonally flooded, are sometimes home to the early-summer flowering summer snowflake or Loddon lily; the summer snowflake is related to the snowdrop.

Bivalves include the pea mussel, while crustaceans include crayfish: sadly the rare native freshwater white-clawed crayfish has all but been replaced by the American signal crayfish that escaped into the river system.

These rivers and streams are also home to a range of birds including ducks, moorhens, coots and swans as well as brightly coloured kingfishers which may be seen sitting patiently on overhanging branches or darting along the river.

From National Nature Reserves to Wildlife Trust Reserves

Within the North Wessex Downs there are a couple of National Nature Reserves, a number of reserves cared for by one of three local wildlife trusts that operate within the region and numerous Sites of Special Scientific Interest (SSSI). There are also six sites – known as Special Areas of Conservation (SAC) – that have been given increased protection to help conserve the world's biodiversity. The sites within the AONB are:

Hackpen Hill (SU352847) – an extensive area of unimproved chalk grassland (including the Devil's Punchbowl and Crowhole Bottom) with a significant population of early gentian, as well as autumn gentian, fragrant orchid, frog orchid, horseshoe vetch, common rock-rose and dwarf thistle.

▲ The picturesque River Lambourn at Eastbury

▲ The sarsen-strewn landscape of Fyfield Down NNR

Little Wittenham Wood (SU572928) – this wooded site includes two ponds that are home to a large number of great crested newts (see also the Earth Trust – page 165).

Kennet Flood Plains (SU313704) – this covers a number of sites that support extensive populations of the minute Desmoulin's whorl snail.

Kennet Valley Alderwoods (SU398675) – an area of broad-leaved deciduous woodland located between Newbury and Marlborough that forms the largest fragment of alder-ash woodland on the Kennet floodplain.

Pewsey Downs (SU106637) – an area of chalk grassland (see National Nature Reserves below) that is home to a significant population of early gentian.

River Lambourn (entire length) – the clear chalk stream, with coarse sediments and submerged plants, provide an ideal habitat for the small, strangely shaped, bullhead fish, which is also known as the 'miller's thumb'.

NATIONAL NATURE RESERVES

Fyfield Down NNR

Situated on a high plateau of chalk grassland to the east of Avebury is Fyfield Down NNR (SU135709), which has the best collection of sarsen stones in Britain and offers a unique combination of geomorphological, biological and archaeological features. The site, which contains around 25,000 sarsen stones in their natural setting, supports rare lichen and moss communities as well as being important for its archaeological interest; part of the reserve is included in the Avebury World Heritage Site. Above Delling Copse there is a sarsen stone or polissoir on which there are several lines formed by our Neolithic ancestors sharpening stone axes on the stone over the centuries.

Pewsey Downs NNR

On the southern edge of the Marlborough Downs is Pewsey Downs NNR (SU108635), a steep south-facing slope including Milk Hill, Walkers Hill and Knap Hill

WILDLIFE

▲ Mute swans at Hungerford Marsh Nature Reserve

overlooking the Vale of Pewsey. This large expanse of flower-rich grasslands has developed over the Upper Chalk mainly because the low level of plant nutrients in the soil helps prevents more vigorous plant species dominating the finer herbs. Plant species found in the reserve include common spotted, frog and fragrant orchids, field fleawort, early gentian, round-headed rampion and bastard toadflax. Butterflies include marble white, skipper, green hairstreak, wall brown and chalkhill blue.

WILDLIFE TRUSTS

The following is a basic summary of what these places have to offer. Full details of all the Wildlife Trust reserves within the North Wessex Downs can be obtained from the relevant trust; please note that some of the reserves have restricted access arrangements in place.

Berkshire, Buckinghamshire and Oxfordshire Wildlife Trust (BBOWT)

Avery's Pightle (SU435651), situated 1¼ miles to the east of Hamstead Marshall, just outside the AONB, is a prime example of one of Berkshire's few remaining unspoiled ancient wet meadows and is known for its abundance of wildlife. The reserve has a rich insect life and nesting bird population and 137 species of plants have been recorded here, including adder's-tongue fern, orchids and, near the stream, a colony of broad-leaved helleborine.

Hungerford Marsh Nature Reserve (SU333687), situated on the banks of the peaceful River Dun at Hungerford, is home to a rich array of wetland birds and wild flowers. Around 120 different bird species have been recorded here, including heron, kingfisher, little grebe, water rail and grasshopper warbler, as well as typical river birds such as mute swans, mallard, moorhen and coot. The reserve is also a good place to see grass snakes and if you are lucky you may catch sight of an otter, especially around dusk or early in the morning.

Inkpen Crocus Field (SU369640) is situated just off Pottery Lane in the little village of Inkpen with the North Hampshire Downs (Combe Gibbet and Walbury Hill) just to the south. In the early twentieth century the banker and naturalist Charles Rothschild (1877–1923) founded the Society for the Promotion of Nature Reserves – the forerunner to the Wildlife Trusts that we have today. The rationale of the society was to both identify and help protect special wildlife sites that became known as Rothschild Reserves. One of these was the Inkpen Crocus Field which is now a designated nature

▲ Purple spring crocus at the Inkpen Crocus Field

reserve owned by the Berkshire, Buckinghamshire and Oxfordshire Wildlife Trust. Here, in late February and early March, tens (even hundreds) of thousands of purple spring crocuses (*crocus vernus*) break into flower, covering the meadow with a mass of purple and white flowers – some striped, others plain.

It's known from village records that the crocuses have been here for over 200 years; however, some believe that they have been there even longer and suggest they were brought back by the Knights Templar during the Crusades.

During the summer and autumn, sheep and a small herd of Dexter cattle are used to graze the grass, keeping the meadow in perfect condition. Following the amazing splash of colour in the early spring, the next best time to visit is the summer, especially around June. At this time the common spotted orchid, heath spotted orchid, meadow saxifrage, cowslip, Devil's-bit scabious, lousewort and twenty-five species of grass can be seen.

Inkpen Common Nature Reserve (SU382643) is a remnant of rare ancient heathland situated close to the Inkpen Crocus Field site. The acid soil of the reserve encourages a variety of heathland plants including gorse, three types of heather, the scarce pale dog-violet and delicate blue heath milkwort; birds to look out for include warblers in late spring and summer.

Kintbury Newt Pond Reserve (SU386663), just to the south of the village, is made up of several ponds and is home to a breeding colony of the nationally rare and protected great crested newt. It also provides ideal nesting conditions for a range of summer migratory birds. The statutory protection of the great crested newts saved the reserve from being developed for housing in the late 1990s. These newts are recognizable by their warty skin and larger size than more common newts such as the smooth and palmate newts that also live here; the male also has a large jagged crest which is especially prominent during its courtship display. Under the oaks at the southern end of the reserve, damp meadow plants such as creeping jenny and wild angelica can be found.

Letcombe Valley (SU377862). The clear waters of the Letcombe Brook that flow along the valley between the villages of Letcombe Bassett and Letcombe Regis are home to some unusual and interesting wildlife including water voles, and fish species including bullheads and brook lampreys – the latter a primitive, jawless fish. Birds include barn owls, kingfishers, herons and little egrets, while at dusk, Daubenton's bats fly across the water's surface catching insects. Wild flowers such as bluebell, common dog violet and wood avens grow amongst the small remnants of ancient woodland, while the fragments of chalk grassland are home to bird's-foot trefoil, field scabious, cowslips and meadow grasses.

Moor Copse (SU633738) is a diverse woodland reserve situated beside the River Pang, just to the south of Tidmarsh, renowned for its woodland flowers, butterflies, moths, dragonflies and damselflies. Plants to look out for include common spotted orchids, twayblades, primrose, early purple orchid, bugle, wood anemone, violet and cowslips and a sea of bluebells in Park Wood in the spring. Birds include the lesser and great spotted woodpecker, while the coppiced areas are good for birds such as blackcap, garden warbler, and chiffchaff. The views of the River Pang as it meanders through the reserve are said to have inspired the illustrator E. H. (Ernest Howard) Shepherd (1879–1976) to draw the

▲ Common spotted orchid

illustrations for the 1931 edition of Kenneth Grahame's classic, *The Wind in the Willows*.

Rack Marsh (SU452692), located at Bagnor, about 2 miles north-west of Newbury, is a good example of an old wet meadow beside the River Lambourn. The reserve is full of rushes and sedges, along with plants such as early marsh-orchids, water avens, bogbean and marsh marigold; birds include blackcap, chiffchaff, grey heron, kingfisher, reed warbler, sedge warbler and whitethroat.

Seven Barrows (SU328828) located 2¼ miles north of Lambourn beside the road to Kingston Lisle is an area of flower-rich chalk grassland surrounding several Bronze Age burial mounds. Over 150 plant species have been recorded here, including horseshoe vetch, chalk milkwort, fragrant orchid, blue harebell and clustered bellflower. The flowers also attract a number of butterflies including the nationally rare marsh fritillary, the chalkhill blue, small blue and the brown argus.

Sole Common Pond (SU411706), with a rich mixture of pond, bog and heath habitats, lies in the middle of ancient common land about 1¼ miles south-west of Wickham. Fifteen species of damselfly and dragonfly have been recorded around the pond, while the bog is home to the insectivorous round-leaved sundew which flowers from June to August; once valued as a herbal remedy for respiratory problems, the sundew secretes a sticky fluid which traps insects that the plant then digests.

Watts Bank (SU331771) is tucked in a fold of the downs just south of Lambourn. This sunny chalk grassland slope provides ideal conditions for wildflowers and at least sixteen different species of grass including quaking-grass. An impressive thirty-two species of butterfly have been recorded here, including chalkhill blue, green hairstreak, marbled white, small copper, brown argus and dingy skipper, while the hazel, blackthorn and bramble scrub provide shelter for many small birds. Flowers include autumn gentian, common spotted orchid, cowslip, Devil's-bit scabious, fairy flax and kidney vetch.

Wiltshire Wildlife Trust

Clouts Wood and Markham Banks (SU135799). Clouts Wood lies just to the south-west of Wroughton in a steep-sided valley which has numerous chalk springs and streams. Parts of the reserve have been wooded since the last Ice Age and there are records of coppicing rights back to the 1600s, making this a very long-standing wooded area. There is a variety of wild flora and fauna present throughout the year. Woodland flowers include wood vetch, nettle-leaved bellflower, bluebells, yellow archangel and the rare Bath asparagus (also known as the spiked star of Bethlehem). Birds include the great spotted woodpecker, treecreeper and the chiffchaff and there are mammals such as foxes, badgers and several species of deer. The adjacent Markham Banks has a variety of chalk grassland wildflowers including common rockrose, horseshoe vetch, spiny restharrow, pyramid orchid and wild thyme.

Ham Hill reserve (SU334616) is situated on the north-facing scarp slope of the North Hampshire Downs alongside the Ham to Buttermere road, with views across the Kennet Valley. A wide range of typical chalk downland plants thrive here, including clustered bellflower, autumn gentian, chalk milkwort, horseshoe vetch and common rockrose. The site is also important for its population of musk orchid, a nationally scarce species that is found in very few sites in Wiltshire; the rare burnt orchid also occurs here in very small numbers. The flowers bring with them a range of butterflies and the reserve is also home to the rather large Roman snail that grows up to 2 inches across, making it Britain's largest snail.

Hat Gate reserve (SU212642) to the north of Burbage is essentially a small section of dismantled railway bought from British Rail in 1988 for £1. The line was part of the Swindon, Marlborough & Andover Railway which closed in 1961 and the site gives an insight to the way that habitat reverts back to a natural state. The site is largely covered in scrub – hazel, hawthorn, blackthorn, dogwood, wayfaring tree, sallow and bramble, with developing woodland – and also some small areas of grassland which supports plants such as eyebright, marjoram, black knapweed, fairy flax, burnet and saxifrage.

High Clear Down (SU235765) is situated in the heart of the Marlborough Downs to the west of Aldbourne. The south-west facing chalk downland slope is noted for its range of downland plant life and in 1998 the Wiltshire Wildlife Trust was able to buy the land with funding from National Power.

Plants found growing on High Clear Down include the early gentian, a small plant with a purple trumpet-shaped flower which blooms during May and June; one of a small number of plants that are found only in Britain. Other plants include common spotted orchid, pyramidal orchid, fragrant orchid, common rockrose and fairy flax. In turn these plants provide food in late spring and summer for a range of butterflies such as the green hairstreak, chalkhill blue, marbled white, duke of burgundy fritillary and brown argus, as well as some day-flying moths such as the cistus forester and wood tiger.

Jones's Mill (SU169611) is a secluded wetland reserve just to the north of Pewsey. The site, which is the only fenland nature reserve in Wiltshire, was previously used as a traditional water meadow that was allowed to flood with mineral-rich, spring-fed waters during the winter to produce an early growth of grass in the spring; the network of water channels can still be seen. Plants such as yellow iris and great horsetail thrive here, as well as rarer plants such as bog pimpernel and flea sedge, while in areas where the vegetation is well grazed the tall spikes of southern marsh orchid and common spotted orchid can be seen.

Morgan's Hill (SU025672) on the western fringes of the North Wessex Downs is one of the highest points in Wiltshire with some lovely views of the surrounding county; the hill got its name when local man John Morgan was executed near the site for the murder of his uncle in the early eighteenth century. The land was part of the Bishop of Salisbury's manor of Bishops Cannings from before 1086 until 1858, when it was sold to the Crown. In the early twentieth century the Marconi Company erected steel masts on Morgan's Hill as part of a chain of wireless stations.

The chalk grassland reserve is designated a Site of Special Scientific Interest (SSSI) for its orchids and butterflies including the endangered marsh fritillary butterfly which feeds on the Devil's-bit scabious found

▲ Small heath butterfly – one of the many types of butterfly that may be seen on chalk grassland

here during late summer. However, there are also a number of important historical features, including a Roman road which once ran between London and Bath and the fifth-century Wansdyke built to defend the northern territory of Wessex.

The reserve is now part of Wiltshire Wildlife Trust's Stepping Stones for Chalk Grassland Project, in partnership with the North Wessex Downs AONB, which has the aim of restoring unmanaged grassland and reverting marginal arable land back to grassland with the use of Herdwick sheep to keep down the scrub and coarse grasses, which would otherwise shade out the wildflowers.

Hampshire and Isle of Wight Wildlife Trust

Old Burghclere Lime Quarry (SU472574). Situated between Beacon Hill and Ladle Hill in the North Hampshire Downs, the steep sides of this disused quarry give shelter to the old workings, making it a good place for butterflies including the small blue whose caterpillars feed on kidney vetch; the quarry floor is also home to fly orchids. (Access is by permit only – contact the Hampshire and Isle of Wight Wildlife Trust.)

More Places to Visit

Beale Park (SU618780), located on the banks of the River Thames just to the north of Pangbourne in Berkshire, is home to a wonderful collection of rare and endangered birds and animals. Gilbert Beale (1868–1967) created Beale Park, which forms part of the Child-Beale Trust, in 1956 by setting aside the large 350-acre site down by the River Thames. In the early days it was little more than a couple of ponds and a large collection of Indian peacocks – some 300 birds by the time of Gilbert's death, aged 99, in 1967. It was Gilbert's great-nephew, Richard Howard, and his family, who made the wildlife park what it is today, specializing in the captive breeding of rare birds and animals.

Bucklebury Common Local Nature Reserve (SU555690) lies to the north-east of Thatcham adjacent to the villages of Upper Bucklebury and Chapel Row. Covering over 860 acres, the common, which is predominantly wooded, also has areas of restored heathland, acid grassland, scrub, wet flushes (where water flows over the surface) and ponds. Parts of the common include ancient semi-natural woodland, but much of the woodland is 'secondary', typically young birch and oak that has reclaimed large parts of the common that were originally heathland as a result of lack of grazing. However, through careful conservation work the characteristic heathland flora has returned and many fauna typical of this habitat, such as the green tiger beetle, slow-worm, adder, nightjar, woodlark and tree pipit are also increasing in number.

A notable feature of the common is The Avenue – a line of several hundred oak trees that were planted along either side of the road heading east from Chapel Row towards Bradfield Southend. The first rows were planted back in the sixteenth century, possibly to commemorate a visit to Bucklebury by Queen Elizabeth I, with a second outer row planted following Wellington's victory at the Battle of Waterloo in 1815.

Bucklebury Farm Park (SU552701), 74 acres in extent, is situated in the Pang Valley. Here you can follow a nature trail, take a wild walk or explore the deer park with its four breeds of deer – red, fallow, Japanese sika and axis – while the farm trail is a great way to learn about farming and farm animal care.

Earth Trust (SU563926), originally the Northmoor Trust, is based at Little Wittenham in the far north-eastern corner of the North Wessex Downs and encompasses the Wittenham Clumps, Little Wittenham Wood Nature Reserve and wildflower meadows.

The distinctive twin, tree-crowned, tops of the Wittenham Clumps – also known as the Sinodun Hills (which includes Brightwell Barrow) – with their panoramic views over South Oxfordshire, are a well-known, and easily recognizable, landmark and have inspired many an artist and poet (see Chapter 6). The Clumps, as they are often called, have been known by several names over the centuries including the 'Berkshire Bubs' (they were originally in Berkshire until boundary changes moved them into Oxfordshire) and

▲ The Loddon lily or summer snowflake

'Mother Dunch's Buttocks'; the Dunch family owned Little Wittenham Manor in the seventeenth century.

The Little Wittenham Wood nature reserve is home to over twenty species of colourful dragonflies and damselflies. Down by the River Thames you might see the club-tailed dragonfly or the banded demoiselle, while the woodland rides are home to the large brown hawker dragonfly at almost 3 inches long. From the hide you might see a kingfisher or be really lucky and catch sight of an otter.

Church Meadow, a wild flower meadow with views across to Day's Lock and beyond to Dorchester Abbey, is home to the rare Loddon lily, also known as the 'summer snowflake', while Broad Arboretum has every species of tree native to Oxfordshire, along with recent introductions such as walnut, sycamore and chestnut. Finally, Neptune Wood was created to commemorate the two-hundredth anniversary of the Battle of Trafalgar. Over 10,000 oak trees were planted here to commemorate the tens of thousands of trees used to construct naval vessels; HMS *Victory*, Nelson's flagship, was built of wood from over 5,000 oak trees.

Incidentally, just to the north, between the River Thames and the historic village of Dorchester on Thames, are two parallel linear earthworks – known as the Dyke Hills – that were constructed as part of an Iron Age settlement. The 'hills' were saved from destruction during the Victorian age by General Pitt Rivers (1827–1900) who helped initiate a national system to protect Ancient Monuments.

Living Rainforest (SU543761) near Hampstead Norreys consists of tropical glasshouses that offer a glimpse of exotic warmer climates. The site originally started life as an orchid nursery before a major conversion resulted in the opening in 1993 of the Wyld Court Rainforest, featuring plants and animals from the world's threatened rainforests. It is now home to hundreds of species of plants, birds, lizards and butterflies along with some of the world's most popular food plants including banana, coffee, cocoa, ginger and vanilla as well as plants that are used for medicines, cosmetics and building materials.

The UK Wolf Conservation Trust (SU590692), based in Beenham, is a non-profit organization founded by the late Roger Palmer in 1995 with the aim of enhancing public awareness and knowledge of wild wolves and their place in the ecosystem. In 1998 the trust imported three European wolves from Eastern Europe and a year later these gave birth to six pups, marking the first birth of European wolves in the UK since they were driven to extinction in the eighteenth century; the last wolf in England is said to have been killed in the fourteenth century. The trust runs a variety of programmes, including walks with wolves, to challenge misconceptions and show that the wolf has a place in wild country. They also support conservation projects around the world aimed at keeping the wolf part of the wild environment for generations to come.

Welford Park (SU409731) is famed for its magnificent display of snowdrops that carpet the woods next to the River Lambourn in early spring – see Stately Homes – page 123.

Useful contacts and information

North Wessex Downs AONB
Denford Manor
Hungerford
Berkshire
RG17 0UN
01488 685440
www.northwessexdowns.org.uk

Tourist Information Centres

Newbury
The Wharf, Newbury, RG14 5AS
01635 30267
www.visitnewbury.org.uk

Oxfordshire
www.visitoxfordandoxfordshire.com

Swindon
Visitor Information Centre
Central Library, Regent Circus
Swindon, SN1 1QG
01793 466454
www.visitwiltshire.co.uk/swindon/tourist-information

Wallingford
Town Hall, Market Place, Wallingford, OX10 0EG
01491 826972
www.wallingford.co.uk

Wantage
Vale and Downland Museum & Visitor Centre
19 Church Street, Wantage, OX12 8BL
01235 771447 or 01235 760176
www.wantage.com

Wiltshire
www.visitwiltshire.co.uk

Ordnance Survey Maps

The North Wessex Downs AONB is located (in parts) on the following maps (the Explorer maps offer much greater detail):

Landranger (1:50,000 or 2 cm per km) series map numbers: 173, 174, 175 and 185

Explorer (1:25,000 or 4 cm per km) series map numbers: 130, 131, 144, 157, 158, 159 and 170

Local Wildlife Trusts

Berkshire, Buckinghamshire and Oxfordshire Wildlife Trust (BBOWT)
The Lodge, 1 Armstrong Road, Littlemore,
Oxford, OX4 4XT
01865 775476
www.bbowt.org.uk

Hampshire and Isle of Wight Wildlife Trust Headquarters
Beechcroft House, Vicarage Lane, Curdridge
SO32 2DP
01489 774400
www.hwt.org.uk

Wiltshire Wildlife Trust Headquarters
Elm Tree Court, Long Street, Devizes
SN10 1NJ
01380 725670
www.wiltshirewildlife.org

Contact Details for Places to Visit

Ardington House
Ardington House, Ardington, OX12 8PY
www.ardingtonhouse.com

Ashdown House and Park
Lambourn, RG17 8RE
01793 762209
www.nationaltrust.org.uk/ashdown-house

Avebury Manor and Gardens; Alexander Keiller Museum
Avebury, SN8 1RF
01672 539250
www.nationaltrust.org.uk/avebury

Basildon House
Lower Basildon, RG8 9NR
0118 9843040
www.nationaltrust.org.uk/basildon-park

Beale Park
The Child Beale Trust, Lower Basildon, RG8 9NW
0844 8261761
www.bealepark.co.uk

Bucklebury Farm Park
Bucklebury, RG7 6RR
0118 9714002
www.buckleburyfarmpark.co.uk

Chiseldon Museum
The Old Chapel, Butts Road, Chiseldon, SN4 0NW
01793 740308
www.chiseldonlhg.org.uk/Museum.htm

Cholsey and Wallingford Railway
5 Hithercroft Road, Wallingford, OX10 9GQ
01491 835067
www.cholsey-wallingford-railway.com

Crofton Beam Engines
Crofton Pumping Station, Crofton, Marlborough, SN8 3DW
01672 870300
www.croftonbeamengines.org

Donnington Castle
Donnington (1 mile north of Newbury, off B4494)
www.english-heritage.org.uk/daysout/properties/donnington-castle

Earth Trust
Little Wittenham, OX14 4QZ
01865 407792
www.earthtrust.org.uk

Englefield House Gardens
Englefield House, Theale, RG7 5DU
www.englefieldestate.co.uk/gardens.html

Highclere Castle
Highclere Park, Newbury, RG20 9RN
01635 253204
www.highclerecastle.co.uk

Kennet and Avon Visitor Centre – Aldermaston
Aldermaston Wharf, Padworth, RG7 4JS
0118 9712868
www.kennetandavonaldermaston.co.uk

Kennet and Avon Visitor Centre and Museum – Devizes
The Kennet & Avon Canal Trust, Couch Lane, Devizes, SN10 1EB
01380 721279
www.kennetandavontrust.co.uk

Kennet Valley at War Museum – Hungerford
Museum is located within Littlecote House
01488 682509
For information visit: www.kennetvalleyatwar.co.uk

The Living Rainforest/Trust for Sustainable Living
Hampstead Norreys, RG18 0TN
01635 202444
www.livingrainforest.org

Ludgershall Castle
Ludgershall (north off A342)
www.english-heritage.org.uk/daysout/properties/ludgershall-castle-and-cross

The Merchant's House
132 High Street, Marlborough, SN8 1HN
01672 511491
www.themerchantshouse.co.uk

Museum of the History of Science
Broad Street, Oxford, OX1 3AZ
01865 277280
www.mhs.ox.ac.uk

Newbury Museum
The Wharf, Newbury, RG14 5AS
01635 519562
www.visitnewbury.org.uk

Pewsey Heritage Centre
Avonside Works, High Street, Pewsey, SN9 5AF
01672 562617
www.pewsey-heritage-centre.org.uk

Reading Museum
Reading Museum & Town Hall, Blagrave Street, Reading, RG1 1QH
0118 9373400
www.readingmuseum.org.uk

Richard Jefferies' Museum
Marlborough Road, Coate, SN3 6AA
01793 466561
www.richardjefferiessociety.co.uk

Sandham Memorial Chapel
Harts Lane, Burghclere, RG20 9JT
01635 278394
www.nationaltrust.org.uk/sandham-memorial-chapel

STEAM – Museum of the Great Western Railway
Fire Fly Avenue, Kemble Drive, Swindon, SN2 2EY
01793 466637
www.steam-museum.org.uk

Tom Brown's School Museum
Broad Street, Uffington, SN7 7RA
www.museum.uffington.net

UK Wolf Conservation Trust
Butlers Farm, Beenham, RG7 5NT
0118 9713330
ukwct.org.uk

Vale and Downland Museum
Church Street, Wantage, OX12 8BL
01235 771447 or 01235 760176
www.wantage.com

Wallingford Museum and Castle
Flint House, 52 High Street, Wallingford, OX10 0DB
01491 835065
www.wallingfordmuseum.org.uk

Welford Park Gardens
Welford, Near Newbury, RG20 8HU
01488 608691
www.welfordpark.co.uk

Wilton Windmill
Wilton Windmill, Wilton, Near Marlborough SN8 3SW
01672 870266
www.wiltonwindmill.co.uk

Wiltshire Heritage Museum
41 Long Street, Devizes, SN10 1NS
01380 727369
www.wiltshireheritage.org.uk

Glossary

Arts and Crafts	Artistic design movement (c.1870–1915) started by the artist and writer William Morris and inspired by the writings of John Ruskin. Evolved into British form of Art Nouveau.
Bailey	Area enclosed by defensive castle walls (see also Motte).
Bronze Age	Period from about 2200–750 BC.
Capital	Top of a column, often decorated.
Chancel	Eastern part of a church for clergy, containing the choir and main altar.
Chancel arch	Arch separating the chancel and nave.
Clerestory	Upper level of nave containing windows to let more light into the church.
Civil War	The English Civil War (1642–1651) was a series of armed conflicts between Parliamentarians (Roundheads) and Royalists (Cavaliers). The Civil War ended with the Parliamentary victory at the Battle of Worcester on 3 September 1651; the wars led to the execution of Charles I and the exile of his son Charles II, although he returned to take the throne in 1660.
Cruciform	Church plan in the shape of a cross, with transepts and central tower, typically Norman or Early Gothic.
Cruck	A large curved wooden beam, part of a pair, used in a building, rising to the apex of a roof.
Decorated	Middle phase of Gothic architecture, characterized by elaborate window tracery and naturalistic carving (c.1280–1380).
Dissolution of the Monasteries	Henry VIII brought about the dismantling of monastic communities and buildings, part of the Reformation in the 1530s.
Domesday Book	Record of the great survey of much of England and parts of Wales completed in 1086, commissioned by William the Conqueror to assess the extent of the land and resources owned at the time.
Doom	Painting of the Last Judgement, usually above the chancel arch of a church.
Early English	First phase of Gothic architecture dominant after Norman, characterized by the earliest pointed arches and simple lancet windows (c.1150–1280).
Flint knapping	When flint is struck it breaks with a shell-like fracture leaving a flat surface; this property can be used to form flint tools such as Neolithic arrow heads or to form flat surfaces for use in house and church walls.
Font	Container, usually stone, sometimes with elaborately carved wooden canopies, which contains holy water for baptism.
Georgian	Architectural style from 1710 to 1830; start of the sash window.
Gothic	Style of architecture which flourished from the late twelfth century until the English Reformation in the 1530s, characterized by the pointed arch, split into three parts: Early English, Decorated, and Perpendicular.
Gothic Revival	Rediscovery of medieval Gothic especially by the Victorians.
Grooved ware	Type of pottery typical of British Neolithic period.
Hall house	Medieval house based on a hall or large room open to the roof.
Iron Age	Period from 750 BC to AD 43.
Jacobean	Early seventeenth-century architecture.
Lancet	Narrow pointed window of the Early English or Early Gothic period.
Lower Cretaceous	Geological period from 145–99 million years ago.
Lychgate	The word 'lych' is Anglo-Saxon for corpse; the gate was used as a resting point during funeral services, where pallbearers could rest while they waited for the priest.
Medieval	Period between the fifth and fifteenth centuries.

Glossary

Mesolithic	Period from 9000–4200 BC.
Middle Ages	Period following the Normans from around 1150 (some include time from Norman Conquest of 1066) to either the Tudor victory at Bosworth (1485) or the start of the English Reformation (1530s).
Motte	A steep-sided man-made mound that would have had a fortification either of wood or stone built on top; the 'motte' was located within a fortified enclosure known as the 'bailey', this design was brought to Britain by the Normans in the eleventh century.
Nave	Main body of the church to the west of the chancel, used by the congregation.
Neolithic	Period from 4200–2200 BC.
Norman	Period from c.1050 to 1200 which followed in the wake of the Norman Conquest of 1066. This period, which is known as Romanesque in Western Europe, brought about the rebuilding of many Anglo-Saxon churches. Churches became more complex and the cruciform floor plan was adopted, with a central tower. This was followed by Transitional architecture (c.1145–90) when both Norman and Gothic styles intermingled.
Palaeogene	Geological period from 65–23 million years ago.
Palaeolithic	Period which started around 11000 BC (sometimes known as Old Stone Age).
Palladian	Style of architecture derived from the sixteenth-century works of Andrea Palladio, which became popular in England in the seventeenth and eighteenth centuries, characterized by elegant Classical proportions, columns and pediments.
Parliamentarians	Often known as Roundheads – supporters of the Parliamentary cause during the English Civil War.
Perpendicular	Final phase of Gothic architecture, characterized by large windows, flattened arches, impressive towers, strong vertical emphasis and fan vaulting (c.1380–1540).
Quaternary	Geological period from 2 million years ago to present.
Quoin	Large dressed corner stone forming external angle on two walls.
RAF	Royal Air Force, formed 1 April 1918 through the amalgamation of the Royal Flying Corps (RFC) and the Royal Naval Air Service (RNAS).
Reformation	In England, Henry VIII severed ties between the English Church and Rome and brought about the adoption of Protestant liturgy.
Roman	Period following the Roman Invasion of AD 43 to AD 410.
Rood	Cross or crucifix bearing the body of Christ, placed between the chancel and nave (see Rood screen).
Rood screen	A carved wooden screen separating the chancel from the nave – mostly destroyed in the Reformation – used to support the figure of Christ crucified (the Rood).
Royal (Coats of) Arms	Coat of Arms of the monarch, usually painted on wood or canvas, to mark their position as Head of the Church of England, placed in churches since the Reformation.
Royalists	Often known as Cavaliers – supporters of the monarchy during the English Civil War.
Saxon	The earliest church architecture that can be seen in some of Britain's oldest churches is Saxon (c.650–1050); fairly simple buildings consisting of a nave and chancel.
Stone Age	Stretching from before 10000 BC to 2000 BC, usually split into three sub-divisions: Palaeolithic Period (up to 10000 BC); Mesolithic Period (10000–5500 BC); Neolithic Period (4000–2000 BC).
Surrealist	Twentieth-century art and literature movement in which unusual or impossible things are shown happening.
Tertiary	Geological period from 65–2 million years ago.
Transept	Part of church at right angles to the nave.
Tree of Jesse	Christ's family tree, representing his descent from Jesse, father of King David, typically depicted in church windows.
Tudor	Period from 1485 to 1603.
Upper Cretaceous	Geological period from 99–65 million years ago.
Victorian	Period from 1837 to 1901.
Whig	British political party with liberal principles from the 1680s to 1850s, later became the Liberal Party.
Winterbourne	Stream that only flows during the wetter winter months of the year.

Further reading

Adkins, Roy and Lesley, and Leitch, Victoria, *The Handbook of British Archaeology* (Constable, revised edn 2008)

Alexander, Steve and Karen, *Crop Circles – Signs, Wonders and Mysteries* (Arcturus Publishing, 2009)

Bailey, David, *Chiseldon Camp from 1914–1922* (Chiseldon Local History Group, 1998)

———*Chiseldon Camp from 1922–1962* (Chiseldon Local History Group, 1998)

de la BédoyŹre, Guy, *Roman Britain: A New History* (Thames and Hudson, 2010)

Berkshire Women's Institute, *The New Berkshire Village Book* (Countryside Books, 1985)

Bettey, J. H., *Wessex from AD1000* (Longman, 1986)

Boyd, David, *The Running Horses: A Brief History of Racing in Berkshire from 1740* (Berkshire County Libraries, 1978)

Burl, Aubrey, *Prehistoric Avebury* (Yale University Press; 2nd revised edn, 2002)

Carrington, Dora and Garnett, David, *Carrington: Letters and Extracts from Her Diaries* (Oxford Paperbacks, 1979)

Christiansen, R., *A Regional History of the Railways of Great Britain: Volume 13 – Thames and Severn* (David and Charles, 1981)

Clarke, Bob, *Prehistoric Wiltshire: An Illustrated Guide* (Amberley Publishing, 2011)

Cobbett, William, *Rural Rides* (Penguin Classics, new edn, 2001)

Cockrell, Peter and Kay, Shirley, *A View from the Hill* (Blewbury Village Society, 2006)

Countess of Carnarvon, *Lady Almina and the Real Downton Abbey: The Lost Legacy of Highclere Castle* (Hodder & Stoughton, 2011)

Cunliffe, Barry, *Britain Begins* (Oxford University Press, 2013)

———*Wessex to AD 1000* (Longman, 1993)

Davies, Hugh, *Roman Roads in Britain* (Shire Publications, 2008)

Davison, Steve, *Walking in the Thames Valley* (Cicerone, 2008)

———*The Ridgeway National Trail* (Cicerone, 2013)

Douglas-Home, James, *Horse Racing in Berkshire* (Sutton Publishing, 1992)

Evans, Sian, *The Manor Reborn: The Transformation of Avebury Manor* (National Trust Books, 2011)

Frere, Sheppard, *Britannia: History of Roman Britain* (Routledge, 1987)

Greenaway, Dick, *Around the Valley of the Pang* (Friends of the Pang, Kennet and Lambourn Valleys, 2007)

Greenaway, Dick and Dunlop, Lesley, *Around the Three Valleys: An Exploration of the Geology, Landscape & History of the Lambourn, Kennet & Pang Valleys in West Berkshire* (Friends of the Pang, Kennet and Lambourn Valleys, 2011)

Grinsell, L. V., *The Archaeology of Wessex* (Methuen, 1958)

Hampshire Women's Institute, *Hampshire Villages* (Countryside Books, 2002)

Hinde, Thomas, *Paths of Progress: a History of Marlborough College* (James and James, 1992)

Hunter, Judith, *A History of Berkshire* (Phillimore, 1995)

Jefferies, Richard, *Wildlife of a Southern County* (Nonsuch Publishing, 2005)

Jenkins, Simon, *A Short History of England* (Profile Books, 2011)

Jenkins, Simon, *England's Thousand Best Churches* (Allen Lane, 1999)

———*England's Thousand Best Houses* (Penguin, 2004)

Leary, Jim and Field, David, *The Story of Silbury Hill* (English Heritage, 2010)

Lee, Alan, *Lambourn: Village of Racing* (Barker, 1982)

FURTHER READING

Maggs, Colin G., *The Branch Lines of Berkshire* (Amberley, 2011)

Marples, Morris, *White Horses and Other Hill Figures* (Sutton Publishing, 1981)

Marshall, Howard, *Reflections on a River* (Littlehampton Book Services, 1967)

Marshman, Michael, *The Wiltshire Village Book* (Countryside Books, 1999)

Morris, John, *The Domesday Book: Berkshire* (Domesday Books – Phillimore, 1979)

——— *The Domesday Book: Hampshire* (Domesday Books – Phillimore, 1982)

Morris, John and Caldwell, C., *The Domesday Book: Oxfordshire* (Domesday Books – Phillimore, 1978)

Morris, John, Thorn, C. and Thorn, F., *The Domesday Book: Wiltshire* (Domesday Books – Phillimore, 1979)

Oakley, Robin, *Valley of the Racehorse* (Headline, 2000)

Oxfordshire Women's Institute, *The New Oxfordshire Village Book* (Countryside Books, 1990)

Pevsner, Nikolaus (founding editor), *The Buildings of England*, series of guides split by county including Wiltshire, Oxfordshire, Berkshire and Hampshire (Yale University Press)

Pollard, Joshua and Reynolds, Andrew, *Avebury: Biography of a Landscape* (The History Press, 2002)

Sands, T. B., *The Didcot, Newbury and Southampton Railway* (Oakwood Press, 1971)

Smith, Esther, *Savernake Forest: The Complete Guide to the Ancient Forest* (Forward Publications, 2010)

———*White Horses of Wiltshire and Uffington* (Forward Publications, 2004)

Stanier, Peter, *Wiltshire in the Age of Steam: C. 1750–1950: A History and Archaeology of Wiltshire Industry* (Halsgrove, 2006)

Stokes, Penelope, *Free Rein: Racing in Berkshire and Beyond 1700–1905* (Penelope Stokes, 2005)

Taylor, Christopher, *Roads and Tracks of Britain* (J. M. Dent and Sons, 1979)

Thomas, Edward, *Richard Jefferies: His Life and Work* (Faber & Faber, 1978)

———*The Icknield Way* (Constable, 1913)

Willis, Steve and Holliss, Barry R., *Military Airfields in the British Isles, 1939–45* (Enthusiasts Publications, 1987)

Wright, Geoffrey N., *Roads and Trackways of Wessex* (Moorland Publishing, 1989)

———*Turnpike Roads* (Shire Publications, 2008)

Yorke, Barbara, *Wessex in the Early Middle Ages* (Leicester University Press, 1995)

Index

(page numbers in italics indicate an illustration)

A
Adam's Grave, 45
Adams, Richard, 153
Ailesbury Column, 137
Aldbourne, 90–91, *91*
Aldbourne Four Barrows, 39
Aldworth, 125
Alfred the Great, King, 29, *29*, 43, 49, 103, 107, 108, 139, 140
Alfred's Castle, 43
Alton Barnes, 126, 140–41
Alton Priors, 126
arable farmland, 156–7
Ardington House, 114, *114*, 168
artists – see Dillon, Anna; Ennion, Eric; Inshaw, David; Nash, Paul; Piper, John; Ravilious, Eric; Spencer, Sir Stanley
Ashampstead, 126–7
Ashdown House, 115, *115*, 168
Ashmansworth, 110
Aston Tirrold, 101
Aston Upthorpe, 101
Atomic Energy Research Establishment – see Harwell Oxford
Aubrey, John, 33, 34, 89
Avebury
 Henge, 33, *33*
 Keiller, Alexander, 32, 34
 Manor, 116–17, *117*, 168
 World Heritage Site, 32–8, *33, 34, 35, 36, 37*
Avington, 127–8

B
Barbury Castle, 44, *44*
Basildon House, 118, 168
Battle of Roundway Down, 50
Beacon Hill, 46
Beale Park, 165, 168
Beckhampton Avenue and Long Barrow, 35
Beenham, 92–3
beer, 61, 89
Berkshire, Buckinghamshire and Oxfordshire Wildlife Trust (BBOWT), 81, 161–3, 167
Betjeman, Sir John, 131, 153
Betjeman, Lady Penelope, 139, *139*
Bincknoll Castle, 44, 49
Binyon, Laurence, 150
Bishops Cannings, 128
Blewburton Hill (fort), *39*, 40
Blewbury, *100*, 101
Bloomsbury Set, 145
Bradfield, *25*, 94–5
breweries, 61
brick-making, 65, 66
Bridges, Robert Seymour, 148
Broad Hinton, 106
Brunel, Isambard Kingdom, 47, 73, 74, 96, 101, 103
Bucklebury, 93
Bucklebury Farm Park, 165, 168
Burbage, 108
Burderop Down, 138
Buttermere, 110

C
Carnarvon, Earls of, 46, 110, 121, 138
causewayed camps, 27, 31, 32, 45 (*see also* Knap Hill, Rybury and Windmill Hill)
Chaddleworth, 105
chalk (including Lower, Middle and Upper), 13–14
 aquifer, 23
 grassland, 157–8
 streams (*see* rivers), 23–6, 158–9
whiting, 65
Cherhill Down, 81, 141
Chesterton, G. K., 143, 145
Chisbury Camp, 45
Chiseldon, 105
 Camp, 51–3, *52, 53*
 Cauldron, 44
 Museum, 168
Cholsey, 101
Christie, Agatha, 101
churches,
 All Saints, Alton Priors, 126
 All Saints, Aston Upthorpe, 110
 All Saints, Burbage, 108
 All Saints, Farnborough, 131, *131*
 All Saints, Hannington, 132, *133*
 All Saints, West Ilsley, 98
 Douai Abbey, Upper Woolhampton, 135–6, *136*
 Holy Cross, Ramsbury, 89–90, *90*, 124
 Holy Trinity, Ardington, 115, *115*, 140
 St Andrew the Apostle, Hurtsbourne Priors, 113
 St Andrew's, Boxford, 129
 St Andrew's, Bradfield, 94–5
 St Andrew's, Chaddleworth, 105
 St Andrew's, Collingbourne Ducis, 108
 St Andrew's, Letcombe Regis, 102
 St Andrew's, Ogbourne St Andrew, 106
 St Augustine of Canterbury, East Hendred, 102
 St Barnabas, Faccombe, 110
 St Bartholomew's, Lower Basildon, 133–4, *134*
 St Clement's, Ashampstead, 126–7, *127*
 St Denys, Stanford Dingley, 135, *135*
 St George's, Ogbourne St George, 106
 St Gregory's, Welford, 124
 St James, Ashmansworth, 110–11, *111*
 St James, Avebury, 127
 St James the Great, Buttermere, 110
 St James the Greater, Eastbury, 130, *130*
 St James the Less, Pangbourne, 96
 St John the Baptist, Mildenhall, 134
 St John the Baptist, Pewsey, 107
 St Laurence's, Tidmarsh, 95, *95*
 St Lawrence, Hungerford, 88, *88*
 St Mark and St Luke, Avington, 51, 127–8
 St Mark's, Englefield, 118–19
 St Martin's Chapel, Chisbury, 109, *109*
 St Mary and St Nicholas, Compton, 98, *99*, 99
 St Mary the Virgin, Alton Barnes, 126
 St Mary the Virgin, Bishops Cannings, 128, *128, 129*
 St Mary the Virgin, Bucklebury, 93, 94, *94*
 St Mary the Virgin, Great Bedwyn, 109, *109*

174

INDEX

St Mary's, Aldworth, 125
St Mary's, Beenham, 92–3, *93*
St Mary's, Cholsey, 101
St Mary's, Collingbourne Kingston, 108
St Mary's, East Ilsley, 98
St Mary's, Great Shefford, 104, *104*
St Mary's, Hampstead Norreys, 132, *132*
St Mary's, Kintbury, 92, *92*
St Mary's, Marlborough, 85
St Mary's, Streatley, 97
St Michael and All Angels, Lambourn, 103, *104*
St Michael and All Angels, Shalbourne, 110
St Michael's, Aldbourne, 90–91, *91*
St Michael's, Aston Tirrold, 101
St Michael's, Blewbury, 101
St Michael's and All Angels, Letcombe Bassett, 102
St Nicholas, Sulham, 95
St Peter and St Paul, Marlborough, 86
St Peter and St Paul, Wantage, 102–3
St Peter and St Paul, Yattendon, *99*, 100
St Peter ad Vincula (or St Peter in Chains), Broad Hinton, 106
St Peter's, Clyffe Pypard, 106
St Peter's, Hurstbourne Tarrant, 111–12
St Peter's, St Mary Bourne, 112–13, *112*
St Swithun's, Wickham, 136–7, *136*, *137*
St Thomas, East Shefford, 129–30, *129*, *130*
Sandham Memorial Chapel, Burghclere, 154, 169
cider, 61–2
Civil War, the, 49–51
Clyffe Pypard, 106
Cobbett, William, 146–7
Collingbourne Ducis, 108
Collingbourne Kingston, 108
Combe Gibbet, 138–9, *138*
Compton, 98–9

Craven (family of), 63, 116
Crofton Pumping Station, 72–3, *73*, 168
crop circles, 144
Cunetio – *see* Mildenhall

D
Devil's Den, 38, *38*
Dillon, Anna, 155
Donnington Castle, 50, *50*, 168
Dragon Hill, 143
Draycot Foliat, 105
Duck, Stephen, 146
Dyke Hills, 166

E
Earth Trust (formerly the Northmoor Trust), 165–6, 168
East Hendred, 101–2, *101*
East Ilsley, 60, 98
East Kennet Long Barrow, 36
East Shefford, 129–30
Eastbury, 130
Englefield House, 118–19, 168
Ennion, Eric, 155
Ermin Way (Roman road), 77

F
Faccombe, 110
Farnborough, 131
film and TV, 56, 61, 91, 99, 117, 118, 119, 121, 139
Finzi, Gerald, 110, *111*
Fosbury Camp, 45–6, *46*
fulling mill, 67–8
Fyfield Down (NNR), 16, *31*, 160, *160*

G
Gault Clay, 12–13, 66
geology, 12–17
GHQ Stop Lines, 53–4
Golding, Sir William Gerald, 153
Grahame, Kenneth, 96, 101, 149–50
Great Bedwyn, 108–9
Great Shefford, 104
Great West Road, 77
Greene, Harry Plunket, 111
grey wethers, 15–16
Grim's Ditch, 40
Grimsbury Castle, 39–40

H
Ham Spray House, 145
Hampshire and Isle of Wight Wildlife Trust, 165, 167
Hampstead Norreys, 132
Hamstead Marshall House, 119–20, *119*
Hannington, 132–3, *132*
Hardy, Thomas, 38, 147–8
Harwell Oxford (formerly Harwell Science and Innovation Campus), 70–71
Hendred House, 101–2
herepath (Green Street), 49
Highclere Castle, 120–21, *120*, *121*, 168
hill forts (*see also* by name), 39–46
history (brief), 27–31
Hocktide (Hungerford), 89
Hughes, Thomas, 143, 147
Hungerford, 86–9
Hurstbourne Priors, 113
Hurstbourne Tarrant, 111

I
Iliffe, Lord, 100, 118
Inshaw, David, 155
iron foundries, 64–5
Ironside, General, 53–4

J
Jefferies, Richard, 138, 149, 169
Jousiffe, Charles William, 64

K
Keiller, Alexander, 32, 34, 116
Kennet and Avon Canal (*see also* GHQ Stop Lines), 71–3, 168
Kintbury, 91–2
Knap Hill Camp, *27*, 45

L
Ladle Hill, 46
Lambourn, 103, *103*
landscape (types of), 17–23
Lansdowne Monument, 139
Lawrence, D. H., 153
Letcombe Bassett, 102
Letcombe Brook, 69, *69*
Letcombe Castle (Segsbury Camp), 40–41, *41*
Letcombe Regis, 102
Liddington Castle, *43*, 44

Littlecote House, 122–3, *122*
Living Rainforest, the, 166, 169
Longstones (Avebury), 35
Lord Wantage (Loyd-Lindsay, Robert), 139–40
Lowbury Hill, 47
Lower Basildon, 133–4
Loyd-Lindsay, Robert (*see* Lord Wantage)
Ludgershall Castle, 50, 169

M
Manger, The, *16*, 143
Marden Henge, 45
Marlborough, 82–6, *82*, *84*, *85*
Merchant's House, 84, *84*, 169
Mound, 83
Marquess of Ailesbury, 123
Medieval castles, 49–51
Membury Camp, 39
Mildenhall, 134
milestones, 77–8
Milk Hill, 17, 48, 144
Mill House (Tidmarsh), 67, 95, 145
mills, 67–8
monuments and memorials (*see also* by name), 137–44
'Moonrakers', 59–60
Morland Brewery (West Ilsley), 61
motte and bailey (castle), 30, 44, 49, 50, 83

N
Nash, Paul, 154
nature reserves, 159–66, *160*
Newbury, 169
Nunhide Tower – *see* Wilder's Folly

O
Ogbourne St Andrew, 105–6
Ogbourne St George, 105–6
Oldbury Castle, 45
Oliver's Castle, 45
Orpheus Mosaic (Littlecote House), 47–8, *47*

P
Pangbourne, 96–7, *96*
Pevsner, Sir Nikolaus, 106
Pewsey, 106–8, 169

175

Pewsey Down (NNR), 160–61
Pilot Hill, 110
Piper, John, 131, *131*, 155
Pitt Rivers, General, 166
Pleydell-Bouverie, Revd Bertrand, 107
Poem Tree (Wittenham Clumps), 147
Ponting, Charles Edwin, 108
Popham, Sir Francis, 122
Poughley Priory, 105
Prior, Matthew, 147

R
racehorses, 62–4
RAF Stations, 54–7
railways, 74–6
Ramsbury, 89–90
Ravilious, Eric, 154
real ale, 61–2
Rennie, John, 71, 72
Ridgeway, the, 40
rivers, *9*, 24–6, *24*, 69, *69*
roads 77–9
Roman, 28, 46–8, *47*, 77
  invasion of Britain, 28
  Orpheus Mosaic, 47–8, *47*
  roads, 77
Rothschild, Charles, 161
Rybury (hill fort), 45

S
St Mary Bourne, 112–13, *112*
Sanctuary, The, 35
sarsen stones, 15–16
Savernake Forest, 123
Saxon earthworks, 48–9
Scott, Sir Walter, 42, 146
Segsbury Camp (Letcombe Castle), 40–41, *41*
Seven Barrows (Lambourn), 43–4

Shalbourne, 109–10
sheep, 60
Silbury Hill, 36–7, *37*
Sinodun Hills (*see* Wittenham Clumps)
smuggling, 59
Sorley, Charles Hamilton, 152–3
Special Area of Conservation, 159–60
Spencer, Sir Stanley, 154
Stanford Dingley, 135
stately homes (*see also* by name), 114–24
Streatley, 97, *97*
Stukeley, William, 33, 34, 38, 83
Sulham, 95
Swanborough Tump, 140
Swift, Jonathan, 102, 146
Sydmonton Court, 111

T
Tan Hill, 17, 48
Tennyson, Alfred Lord, 126
Thomas, Edward, 130, 151–2
Tidmarsh, 95
Tottenham House, 123
tourism, 79–81
Townsend, Joseph, 107
Tubb, Joseph, 147
Tull, Jethro, 60, 110
Turnpike Trusts, 77–9
Tutti Day (*see* Hocktide)

U
Uffington
  Castle, 41–2
  Tom Brown's School Museum, 143, 169
  White Horse, 41–2, *41*
Upper Greensand, 12–13
Upper Woolhampton, 135–6, *135*

V
Vale and Downland Museum (Wantage), 103, 169
Vale of Pewsey, 106–8
vineyards, 62

W
Walbury Hill, 46, 58, *58*
Wallingford Castle, 49, 169
Walwyn, Fulke, 64
Wansdyke, 48–9, *48*
Wantage, 102–3
water mills, 67–8
watercress, 68–70
Waterhouse, Alfred, 100
Watership Down, 153
Wayland's Smithy, 42–3, *42*
Welford Park, 123, 169
West Ilsley, 98
West Kennet Avenue, 34–5, *34*
West Kennet Long Barrow, 35–6, *35*, *36*
Whistler, Sir Laurence (windows), 111, 130, *130*, 132–3, *133*, 151
white horses
  Alton Barnes, 140–41, *140*
  Broad Town, 141
  Cherhill, 141, *141*
  Devizes, 142
  Hackpen, 142, *142*
  Ham Hill (Inkpen), 143
  Marlborough, 142
  Pewsey, 142
  Rockley, 143
  Uffington, 41–2, *41*, 143
whiting, 65
Wickham, 136–7
Wilder's Folly, 140, *140*
William of Orange, Prince (William III), 87

Williams, Alfred, 138, 150–51
Wilton Water, 73
Wilton Windmill, 68, *68*, 169
Wiltshire Heritage Museum (Devizes), 32, 169
Wiltshire Wildlife Trust, 163–5, 167
Winchcombe, John (aka Jack of Newbury), 30, 68, 93
windmill, 68
Windmill Hill, 32
wine, 61–2
Winter, Fred, 64
Wittenham Clumps, 40
Wolf Conservation Trust (UK), 166, 169
Wolsey, Thomas, *85*, 86, 105
woodland, 158
World Wars (First and Second), 51–8
writers – *see* Adams, Richard; Betjeman, Sir John; Binyon, Laurence; Bridges, Robert Seymour; Chesterton, G. K.; Christie, Agatha; Cobbett, William; Duck, Stephen; Golding, Sir William Gerald; Grahame, Kenneth; Hardy, Thomas; Hughes, Thomas; Jefferies, Richard; Lawrence, D. H.; Prior, Matthew; Scott, Sir Walter; Swift, Jonathan; Thomas, Edward; Tubb, Joseph; Williams, Alfred

Y
Yattendon, 100